BestMasters

Mit **„BestMasters“** zeichnet Springer die besten Masterarbeiten aus, die an renommierten Hochschulen in Deutschland, Österreich und der Schweiz entstanden sind. Die mit Höchstnote ausgezeichneten Arbeiten wurden durch Gutachter zur Veröffentlichung empfohlen und behandeln aktuelle Themen aus unterschiedlichen Fachgebieten der Naturwissenschaften, Psychologie, Technik und Wirtschaftswissenschaften. Die Reihe wendet sich an Praktiker und Wissenschaftler gleichermaßen und soll insbesondere auch Nachwuchswissenschaftlern Orientierung geben.

Springer awards **“BestMasters”** to the best master’s theses which have been completed at renowned Universities in Germany, Austria, and Switzerland. The studies received highest marks and were recommended for publication by supervisors. They address current issues from various fields of research in natural sciences, psychology, technology, and economics. The series addresses practitioners as well as scientists and, in particular, offers guidance for early stage researchers.

Roman Delorme

Optimization of Transformation Pathways for the Residential Building Sector

Springer Vieweg

Roman Delorme
RWTH Aachen University
Aachen, Germany

ISSN 2625-3577 ISSN 2625-3615 (electronic)
BestMasters
ISBN 978-3-658-47465-2 ISBN 978-3-658-47466-9 (eBook)
https://doi.org/10.1007/978-3-658-47466-9

This Springer Vieweg imprint is published by the registered company Springer Fachmedien Wiesbaden GmbH, part of Springer Nature.
The registered company address is: Abraham-Lincoln-Str. 46, 65189 Wiesbaden, Germany

Foreword

The European countries, and Germany in particular, have set goals to reach climate neutrality by the mid of this century. Many sectors contribute to these goals, and the building sector is an important one of them. Latest standards do not only have to be respected in newly erected buildings. The existing stock must undergo renovation or reconstruction as well. Technological options are plenty, some more suited than others, some more effective than others, depending on the building type, the region, the usage, in the end of course also depending on the budget and available resources. Finding a best option for a single building may seem comparatively easy. However, for larger housing companies, with tens of thousands of houses to maintain, many far-reaching decisions have to be made simultaneously, impacting a time horizon of potentially two decades: Which specific technology should be applied, in what dimension, to which houses, in which years? Not only money is a scarce resource. Engineers, consultants, administrators are needed, and of course the material, too. Everything is connected to everything.

We quickly see that this is not only a huge technical task, it is first and foremost a huge planning task. Decision support is sorely needed.

Huge planning tasks may seem overwhelming and risky because we do not know where to start, how to handle the many interdependencies and uncertainties, the countless restrictions and rules. A classical approach might be to break the huge task down into manageable parts. But we don't necessarily have to. There is a technology that helps us tackle the planning task in its entirety. This technology is mathematics. The field of mathematical optimization gives us models, a language to describe a planning task with all its decisions, constraints, and goals in such a way, that every potential plan is respected. This sounds totally unimaginable, but this is exactly what it is, the consideration of all options. Even better, mathematical optimization offers powerful algorithms that compute the best possible solutions to these models which can help support the decision makers in concrete, directly interpretable, actionable ways. When designed together by mathematicians and engineers, this kind of decision support can fully exploit the potential in a planning task. And, as it stands, we will have to exploit every little bit of potential when we want to reach the ambitious climate goals.

Transition is a big word. I believe that the word is adequate in the light of the magnitude of the task. I am convinced that the importance and impact of a planning task must be matched by the planning technology that we employ to resolve the task. Mathematical optimization offers such a planning technology, but it is not so well known, and even less applied. In that sense, I hope that a thesis like this one contributes also to the transition in our heads. The engineer transforms to a mathematically supported engineer. This change comes with pains and with opportunities. I am indebted to Roman Delorme for the ability and the courage to have embarked on this journey to impressively demonstrate the potential of this planning transition. I would love to see much more of this!

Aachen
December 2024

Marco Lübbecke

Abstract

The mathematical optimization of transformation pathways for the residential building sector enables deriving strategies to cost-efficiently mitigate greenhouse gas emissions in the building sector. To that end, transformation pathways should simultaneously provide information about *which action* to take in *which building* and in *which year*. However, only a few studies attempt to develop optimization frameworks that address these three questions simultaneously. Among the studies that incorporate all three questions, some predetermine pathways for single buildings, others only consider small building portfolios or the optimization framework does not provide proof of optimality. Therefore, optimization frameworks are required that facilitate the optimization of transformation pathways.

In this thesis, we define the problem of determining optimal transformation pathways for a residential building stock. We identify three logically and temporally interconnected levels of the problem. On the first level, different restricted resources need to be allocated over time and across the building stock. On the second level, the allocated resources determine the individual building energy system design and building efficiency for each building and year. Lastly, on the third level, depending on the building efficiency, building operations need to always satisfy energy demands with the current building energy system design. We propose a MILP model for the introduced problem. The three parts of the MILP model are only linked by sets of variables, rendering Benders decomposition a viable approach. Therefore, based on the different levels of the problem, we formulate various Benders decompositions of the original MILP model and integrate them into the branch-and-cut process of the commercial MILP solver Gurobi. We then introduce constraint modifications for the different decompositions aiming to reduce the time required for Benders cut generation and valid inequalities to prevent cases of infeasibility for primal Benders subproblems. Additionally, we introduce a strengthening technique for Benders optimality cuts of decompositions that use building efficiency states as linking variables. We implement a construction heuristic that employs a simplified version of the original MILP model with cost estimates to produce feasible solutions for the original model. Moreover, we implement an in-out method to improve the dual bound and reduce the number of used Benders cuts.

To evaluate our implementation, we construct different instance sets that correspond to different building stock sizes with respectively 10, 20, and 50 buildings and an optimization time horizon of 20 years. None of the instances is solved to optimality within the given time limit by any considered optimization strategy. We consistently find the smallest relative MILP gaps at the time limit for solution strategies employing a decomposition that only includes constraints of building energy system operations in Benders subproblems. Importantly, when Benders subproblems and Benders cuts of this decomposition are further disaggregated on a yearly basis, relative MILP gaps at the time limit are further reduced. For strategies based on this decomposition, the modification of constraints in Benders subproblems reduces time in callbacks but leads to higher relative MILP gaps at the time limit, probably due to the dual degeneracy of Benders subproblem solutions. The construction heuristic reliably determines high-quality primal bounds. We observe that the proposed constraint modifications reduce time spent in solver callbacks but do not consistently reduce relative MILP gaps. In a heuristic strategy refinement, we then iteratively improve our solution strategy. The solution strategy that most reliably yields the smallest relative MILP gaps at the time limits employs the aforementioned decomposition with further disaggregated Benders subproblems and cuts, no modified constraints, and the in-out method. Importantly, when optimizing large instances with 50 buildings over a time horizon of 20 years, we find that our best-performing strategy reaches relative MILP gaps at the time limit of approximately 5 % while Gurobi attains relative MILP gaps of 20 % and 10 % if Gurobi performs a warm start with our construction heuristic.

Contents

Abbreviations

BAT	Battery
B&B	Branch-and-bound
B&C	Branch-and-cut
BO	Building operations
BOEP	Building operations and efficiency pathways
BOESEP	Building operations, energy system and efficiency pathways
BOESP	Building operations and energy system pathways
BOI	Boiler
CHP	Combined heat and power
CO_2eq	CO_2-equivalent
DHW	Domestic hot water
DOD	Depth-of-discharge
EH	Electrical heater
GA	Genetic algorithm
GWP	Global warming potential
HP	Heat pump
LHS	Left-hand-side
LP	Linear programming
MILP	Mixed-integer linear programming
NPV	Net present value
O&M	Operations and maintenance
PV	Photovoltaic modules
RHS	Right-hand side
RMP	Relaxed master problem
SOC	State-of-charge
SP	Subproblem

STC	Solar thermal collector
TES	Thermal energy storage
TGW	Total global warming
TRY	Test reference year

List of Figures

List of Tables

Introduction

1

In Germany, residential buildings accounted for approximately 27 % of overall CO_2-equivalent (CO_2eq) greenhouse gas (GHG) emissions in 2020 (Destatis, 2022; UBA, 2022). Due to its significant contribution to overall GHG emissions, the residential building sector plays a crucial role in achieving net neutrality of CO_2eq GHG emissions by 2045 as determined by the German Federal Climate Change Act (BMJ, 2019). However, in 2019, approximately 35.9 % of the German residential building stock was completely unrenovated, with 51.4 % being only partially renovated (UBA, 2019). Furthermore, the current renovation rate stagnates at around 0.9 % a^{-1} and must be increased to approximately 1.9 % a^{-1} to meet respective sector goals by 2045 (dena, 2021). The limited availability of resources such as building energy system devices and workforce poses a bottleneck for further increasing the renovation rate (Lutz et al., 2018; Bazejczak and Edler, 2021). To meet climate targets of the residential building sector, it is therefore important to increase the availability of limited resources and to make efficient use of them.

The purpose of building energy systems is to satisfy building energy demands at all times. Importantly, the operation of a building energy system is characterized by the dynamic interdependence of the energy *supply* and *demand* sides of the building. Since the supply side of buildings is represented by the building energy system design and the demand side depends on currently realized building efficiency measures, changes to the building energy system design or the realization of new efficiency measures impact building energy system operations and thus also entail additional costs and GHG emissions connected to building energy system operations. Furthermore, changes to the building energy system design or the realization of efficiency measures are connected to investment costs or embodied GHG emissions. *Transformation pathways* of the residential building sector describe in what particular way resources are used to transform building energy systems or efficiency measures over a defined time horizon for a given building stock. In particular,

R. Delorme, *Optimization of Transformation Pathways for the Residential Building Sector*, BestMasters, https://doi.org/10.1007/978-3-658-47466-9_1

transformation pathways precisely state *which action* to perform in *which building* and *which year* (Pannier et al., 2021). Transformation pathways, therefore, define valid allocations of resources, e.g., building energy system devices and workforce, such that limitations in resources are respected and building energy systems can be operated such that building energy demands are satisfied. Due to the heterogeneity of buildings in residential building stocks with respect to their current building energy system designs and existing building efficiency measures, different allocations of resources and thus transformation pathways may lead to significant differences concerning ecological or economic objectives.

Mathematical optimization provides tools to determine transformation pathways that are optimal with respect to an economic or ecological objective. Optimal transformation pathways are useful as they provide insights to policymakers and thus help drive the transformation of the residential building sector more efficiently. In the context of such decision-related problems, mixed-integer linear programming (MILP) models enable describing such problems mathematically. However, depending on the size and complexity of the considered MILP models, solving them may become computationally intractable. For certain model structures, MILP models can be decomposed into several other problems that are collectively easier to solve. In the presence of linking variables that couple different sets of constraints of the original model, *Benders decomposition* is a model decomposition technique that may be particularly suited (Benders, 1962). Benders decomposition is a method based on a sequence of projection, outer linearization, and relaxation (Geoffrion, 1970). To apply the Benders decomposition method, the original MILP model is decomposed into a so-called relaxed Benders master problem (RMP) and several Benders subproblems (SPs). First proposed by Benders (1962), various applications, acceleration methods, and generalization techniques have been proposed (Rahmaniani et al., 2017).

1.1 Related Work

In the context of single-building optimization, numerous optimization frameworks have been developed (Evins, 2013). Those studies generally differ greatly in the way that they consider the interdependence of building energy system design, efficiency measures, and building energy system operations as well as their respective objective. Only in recent years, research has focused more on mathematical optimization in the context of building portfolios or building stocks (Gabrielli and Ruggeri, 2019). Pannier et al. (2021) conduct a literature review on recent studies that address setting action priorities for refurbishments at building stock level and answer at least one of the questions *which action*, *which building* and *which year*. The authors find that in most of their reviewed studies, not all three questions are considered simultaneously. Furthermore, some of the reviewed studies consider building portfolios of real estate property owners with yearly budget constraints instead of buildings stocks. In the following, we briefly review selected works that aim to answer at least two of the aforementioned questions in the context of mathematical optimization. Here, we also consider works that are not considered by Pannier et al. (2021) or that are more recent than their review. For a broader review of recent literature in this field, we refer to the reviews of Pannier et al. (2021) and Ruggeri et al. (2020).

Most studies that simultaneously consider the questions *which action*, *which building* and *which year* in their optimization framework rely on choosing from a selection of different retrofit plans for each building. Gabrielli and Ruggeri (2019) develop a two-stage decision support tool for which solutions provide an answer to the questions *which action* and *which building*. The authors cluster a building stock with a k-means algorithm and define sets of retrofit plans, using the same set of retrofit plans for each building of the same cluster. Subsequently, the authors solve an integer problem (IP), in which for each building exactly one retrofit plan must be chosen while respecting a maximum budget. Furthermore, the authors use a weighted objective function, considering net savings, environmental impact, internal comfort, and restoration aspects. Similarly, Taillandier et al. (2017) propose a multidimensional knapsack problem with conflicts for the optimization of maintenance schedules for a real estate property owner. The authors consider a set of retrofit actions where each retrofit action can be performed at most once during the optimization time horizon. Each retrofit action is given a cost factor and for each year of the optimization time horizon, a given budget must not be exceeded. In their objective function, Taillandier et al. (2017) consider regulatory compliance, asset value, and client satisfaction of each retrofit action. Each retrofit action concerns a specific maintenance action on a component of a specific building. The set of possible retrofit actions is proposed by a pool of technical experts related to a French company that is considered for a case study in Taillandier et al. (2017). Lastly, Petkov et al. (2023) propose a two-stage optimization framework to simultaneously optimize building energy system design and building efficiency measures for buildings of a building portfolio owner. First, the authors utilize a MILP model developed in Petkov et al. (2022) to conduct various multi-objective cost and life cycle CO_2 optimizations for each separate building in a given portfolio. Subsequently, Petkov et al. (2023) use an IP to determine optimal retrofit plans concerning total portfolio costs and total portfolio CO_2 emissions. In their IP, they assign exactly one of the previously determined solutions to each building while ensuring that yearly cost and total CO_2 budgets for the building portfolio are not exceeded. In all aforementioned studies, a set of retrofit plans needs to be determined before assigning retrofit plans to buildings of the considered building portfolios. Moreover, while the approaches of Gabrielli and Ruggeri (2019) and Taillandier et al. (2017) require expert knowledge for the definition of retrofit plans, all three studies disregard the interdependence between retrofit plans themselves and different resource restrictions.

Conversely, in the work of Pannier et al. (2021), a genetic algorithm (GA) is used to optimize retrofit plans for multiple buildings of a building portfolio owner in the greater Paris area. For each building, a set of available building efficiency measures is defined as well as a set of predefined building energy system structures. The GA then optimizes total investment costs as well as the final energy consumption for heating by iteratively improving an initial solution. Notably, the optimization framework of Pannier et al. (2021) considers multiple buildings as well as multiple allocation decisions along a time axis, effectively answering all three questions *which action, which building* and *which year*. Furthermore, yearly and total investments are constrained, therefore coupling decisions across buildings. However, the GA approach of Pannier et al. (2021) does not provide a certificate of optimality. Furthermore, the approach of Pannier et al. (2021) focuses heavily on building efficiency measures, while only considering the replacement by a new gas boiler or the connection to district heating on the supply side with building energy system devices.

In some studies, more sophisticated MILP models are developed that, however, do give rise to solutions that dot not answer all questions *which action, which building*, and *which year*. Richarz et al. (2022) develop a MILP model that simultaneously optimizes the building energy system design, the realization of building efficiency measures as well as building operations for a non-residential building over multiple years. Devices constituting the building energy system as well as building efficiency measures can be chosen from a predefined superstructure. The authors then maximize the net present value resulting from investment decisions and operations of the building energy system. The approach of Richarz et al. (2022) therefore determines investment decisions into devices and efficiency measures for a single building over multiple years but does not consider multiple buildings simultaneously. Lastly, Kotzur et al. (2020) develop a MILP model to optimize the building energy system design, building efficiency measures, and building operations of multiple archetype buildings of a building stock in a bottom-up optimization framework. In their approach, Kotzur et al. (2020) first aggregate a residential building stock using an optimization-based clustering algorithm. The authors then develop a MILP model that simultaneously considers the building energy system, building efficiency measures, and building operations of a single building. Then, for each archetype building, a single MILP instance is solved and the respective solutions are aggregated for all archetypes of the building stock. While the optimization framework of Kotzur et al. (2020) is applied to multiple buildings, it does not consider restrictions in resources and resulting interdependencies. Furthermore, their model only parametrically considers the time horizon while investment decisions may only be done at a singular point in time.

1.2 Contributions

While most studies do not consider the three questions *which action*, *which building* and *which year* simultaneously, e.g., Richarz et al. (2022) and Kotzur et al. (2020), others employ a two-step framework in which retrofit plans are predetermined before being assigned to different buildings of a building stock or portfolio (Gabrielli and Ruggeri, 2019; Taillandier et al., 2017; Petkov et al., 2023) or rely on optimization methods that do not prove optimality (Pannier et al., 2021). For this reason, we identify the need for an optimization framework that holistically considers all three questions for the determination of optimal transformation pathways. While MILP models can be used to describe such problems, solving them is computationally hard due to their size and model complexity. To facilitate the optimization of such a model, we develop a solution approach based on Benders decomposition to determine optimal transformation pathways for a residential building stock. In particular, our contributions are the following:

1. We formally define the problem of determining optimal transformation pathways for a residential building stock and develop a corresponding MILP model.
2. We propose different Benders decompositions for the developed MILP model along with constraint modifications that aim to accelerate the cut generation process for Benders decomposition and a strengthening technique for Benders optimality cuts.
3. We develop and implement a solution approach that integrates Benders decomposition into the branch-and-cut (B&C) process of the commercial MILP solver Gurobi (Gurobi Optimization, LLC, 2023). Furthermore, we employ various acceleration techniques and implement a tailored construction heuristic to warm start the optimization process.
4. We introduce different instance test set to benchmark our solution approach and validate the different acceleration techniques.

The contribution and focus of this thesis are depicted in Fig. 1.1. Note that, for this thesis, we develop an optimization framework to determine optimal transformation pathways for instances where we estimate several parameters. Therefore, the study conducted here mainly is a feasibility study that investigates computational possibilities to determine optimal transformation pathways for the described problem. In a future application of the developed optimization framework, more accurate data on the residential building stock and available resources first would need to be acquired and aggregated by appropriate methods to yield representative building archetypes and their parametrizations. The optimization framework developed

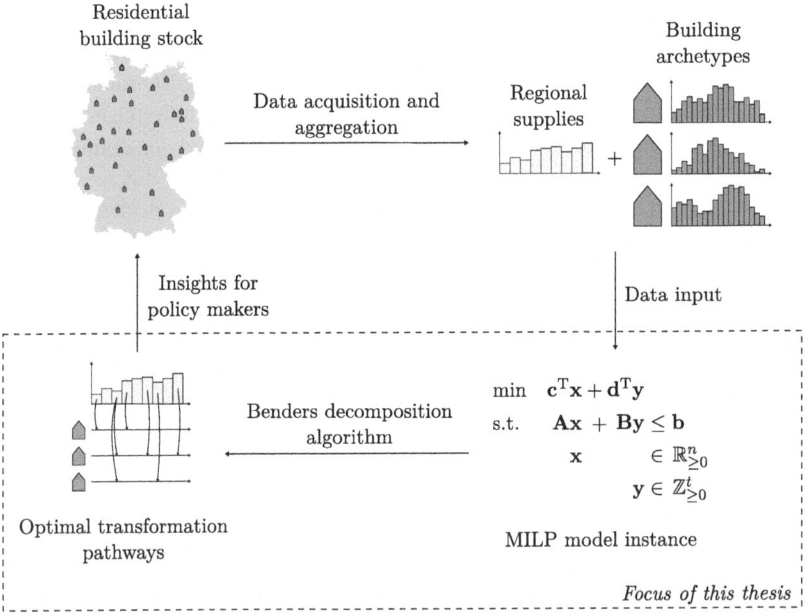

Fig. 1.1 Determining optimal transformation pathways for a residential building stock using building stock aggregation and employing a MILP optimization framework based on Benders decomposition. Using appropriate data acquisition and aggregation techniques, the residential building stock is aggregated into building archetypes, and regional supplies for building energy system devices and workforce are identified. Parameters of building archetypes and supplies are then used as inputs for the developed MILP model that determines optimal transformation pathways for the set of considered archetypes of buildings. The MILP model is solved using a solution approach that employs Benders decompositions along with various acceleration methods. Optimal solutions of the proposed model can eventually be used to derive insights for policymakers. This thesis focuses on the development of the MILP model and the corresponding solution approach based on Benders decomposition

in this work can then be used to determine optimal transformation pathways for the residential building stock. Optimal transformation pathways could eventually be used to provide insights to policymakers and help increase the efficiency of the transformation process of the residential building stock.

1.3 Structure

This thesis is organized as follows: In Chap. 2, we introduce the concept of linear optimization for the continuous and mixed-integer case and discuss fundamental aspects of Benders decomposition as well as acceleration techniques of Benders decomposition algorithms relevant to our work. In Chap. 3, we formally define the problem and present a MILP model for the given problem. Chapter 4 introduces optional constraint modifications of the original model that aim to reduce the time needed for Benders cut generation as well as different Benders decompositions that we derive for the developed model. Furthermore, we present a strengthening technique for Benders optimality cuts that can be applied in some of the introduced decompositions. Our solution approach and implementation aspects are presented in Chap. 5. The method to generate instances used to test the developed method is introduced and the obtained results are presented and discussed in Chap. 6. Finally, the thesis concludes in Chap. 7, where key aspects are summarized, notable findings are highlighted, and potential avenues for future improvements are suggested.

2 Theoretical Background

In this thesis, we develop a model and a solution approach for the optimization of transformation pathways for the residential building sector. This chapter introduces the necessary theoretical concepts for constructing the optimization model and implementing the solution approach.

Section 2.1 presents relevant concepts of linear optimization, including duality and the simplex method. Subsequently, we introduce mixed-integer optimization problems in Sect. 2.2 and review solution methods for this problem class. Lastly, in Sect. 2.3, we introduce the Benders decomposition algorithm, a technique for solving large-scale mixed-integer optimization problems with linking variables. Additionally, we explore different acceleration methods that can enhance the efficiency and performance of the Benders decomposition algorithm.

2.1 Linear Programming

In *Linear Programming* (LP), one minimizes or maximizes a linear functional over a polyhedron (Schrijver, 2011). Without loss of generality, we here define the problem (LP) by

$$
\text{(LP)} \qquad
\begin{aligned}
z^{\mathrm{LP}} = \min \quad & \mathbf{c}^{\mathrm{T}}\mathbf{x} && (2.1)\\
\text{s.t.} \quad & \mathbf{A}\mathbf{x} \leq \mathbf{b} && (2.2)\\
& \mathbf{x} \in \mathbb{R}^{n}_{\geq 0} && (2.3)
\end{aligned}
$$

where $\mathbf{x} \in \mathbb{R}^{n}_{\geq 0}$ is a vector of non-negative continuous decision variables with objective coefficients $\mathbf{c} \in \mathbb{R}^{n}$. The feasible region of the problem is a polyhedron given by the inequality system $\mathbf{A}\mathbf{x} \leq \mathbf{b}$, where $\mathbf{A} \in \mathbb{R}^{m \times n}$ and $\mathbf{b} \in \mathbb{R}^{m}$. In this

R. Delorme, *Optimization of Transformation Pathways for the Residential Building Sector*, BestMasters, https://doi.org/10.1007/978-3-658-47466-9_2

context, we call $\mathbf{b}$ the *right-hand side* (RHS) of the inequality system in (LP). Problem (LP) either has an optimal solution, it is infeasible or unbounded. If (LP) has an optimal solution, an optimal solution $\mathbf{x}^*$ is always attained at a vertex of the polyhedron given by $\mathbf{Ax} \leq \mathbf{b}$ (Chvátal, 1983).

2.1.1 Duality

For a given problem (LP), one can always construct the so-called *dual problem* (DLP) of (LP) with

$$\text{(DLP)} \qquad \begin{aligned} z^{\text{DLP}} = \max \quad & \boldsymbol{\pi}^{\mathrm{T}}\mathbf{b} && (2.4)\\ \text{s.t.} \quad & \boldsymbol{\pi}^{\mathrm{T}}\mathbf{A} \leq \mathbf{c}^{\mathrm{T}} && (2.5)\\ & \boldsymbol{\pi} \in \mathbb{R}^m_{\leq 0} && (2.6) \end{aligned}$$

that is also an LP problem. In the context of duality, we henceforth call (LP) the *primal problem* to highlight its difference to (DLP). In (DLP), $\boldsymbol{\pi} \in \mathbb{R}^m_{\leq 0}$ are non-positive continuous decision variables. While dual variables $\boldsymbol{\pi}$ are in one-to-one correspondence with primal constraints $\mathbf{Ax} \leq \mathbf{b}$, dual constraints $\boldsymbol{\pi}^{\mathrm{T}}\mathbf{A} \leq \mathbf{c}^{\mathrm{T}}$ are in one-to-one correspondence with primal variables $\mathbf{x}$ (Chvátal, 1983). Furthermore, the RHS of the primal problem $\mathbf{b}$ becomes the vector of objective coefficients in the dual and vice versa. Lastly, while we minimize the objective function in primal problem (LP), the objective in (DSP) is maximized.

In the following theorems, we introduce some theoretical concepts that connect both problems (LP) and (DLP):

Theorem 1 (Strong Duality. Proof in Nemhauser and Wolsey (1988).). *If* z^{LP} *or* z^{DLP} *is finite, then both* (LP) *and* (DLP) *have finite optimal value such that*

$$z^{\text{LP}} = \mathbf{c}^{\mathrm{T}}\mathbf{x}^* = \boldsymbol{\pi}^{*\mathrm{T}}\mathbf{b} = z^{\text{DLP}}. \qquad (2.7)$$

Theorem 2 (Farkas Lemma. Proof in Nemhauser and Wolsey (1988).). *Let* $\mathbf{A} \in \mathbb{R}^{m\times n}$ *and let* $\mathbf{b} \in \mathbb{R}^m$. *Then either* $\{\mathbf{x} \in \mathbb{R}^n_{\geq 0} : \mathbf{Ax} \leq \mathbf{b}\} \neq \emptyset$ *or (exclusively) there exists* $\boldsymbol{\pi} \in \mathbb{R}^m_{\leq 0}$ *such that* $\boldsymbol{\pi}^{\mathrm{T}}\mathbf{A} \leq \mathbf{0}^{\mathrm{T}}$ *and* $\boldsymbol{\pi}^{\mathrm{T}}\mathbf{b} > 0$.

The concept of strong duality introduced in Theorem 1 is important since it guarantees that if only interested in the optimal objective value z^{LP}, we can fall back to the optimal objective value z^{DLP} of (DLP). In some way, Theorem (1) therefore establishes some interchangeability of problems (LP) and (DLP), depending on the context. Farkas Lemma, introduced in Theorem 2, states that exactly when problem (LP) is infeasible, there exists a dual ray $\boldsymbol{\pi} \in \mathbb{R}^m_{\leq 0}$, i.e., a direction of unboundedness of (DLP), that proves infeasibility of (LP).

2.1.2 Simplex Method

The simplex method makes use of the fact that if an optimal solution for problem (LP) exists, such an optimal solution may always be attained at a vertex of the polyhedron $\mathbf{Ax} \leq \mathbf{b}$. Standard implementations of the simplex method generally make use of two distinct phases. In phase one, the simplex method computes an initial feasible solution, i.e., a vertex of the polyhedron $\mathbf{Ax} \leq \mathbf{b}$, or yields that (LP) is infeasible. In the second phase, the simplex method then either discovers an optimal solution by iteratively improving the current feasible solution or yields that (LP) is unbounded (Chvátal, 1983). The solution is iteratively improved by moving to neighboring vertices with improving objective values. A full description of the simplex method and different implementation aspects can be found in Chvátal (1983) and Schrijver (2011). Importantly, in every iteration in phase two of the simplex method, a corresponding dual solution is automatically computed. Therefore, by Theorem 1, if the simplex method terminates with an optimal solution $\mathbf{x}^*$, it also discovers an optimal dual solution $\boldsymbol{\pi}^*$. This is an important aspect of the simplex method as it allows to also compute an optimal dual solution $\boldsymbol{\pi}^*$ of (DLP) by solving the primal (LP).

The previously described method corresponds to the primal simplex method. Next to the primal simplex method, a dual simplex method can be used that can be considered as the primal simplex method applied to the dual problem (DLP), in an alternative administration (Schrijver, 2011). If multiple processors are available, LP solvers like Gurobi can employ concurrent solving techniques that solve problem (LP) concurrently employing the primal and dual simplex method or other algorithms like the barrier method (Gurobi Optimization, LLC, 2023).

Significantly, given a primal feasible solution $\mathbf{x}$ or a dual feasible solution $\boldsymbol{\pi}$, the primal or dual simplex method can respectively be started directly from phase two. In case the objective function of (LP) is modified, the feasible region corresponding to the polyhedron in (LP) remains unchanged. Therefore, if (LP) has been optimized before and a previously optimal primal solution $\mathbf{x}^*$ is available, the same vector $\mathbf{x}^*$ is a feasible solution for the now modified (LP). Therefore, the simplex method can be started in phase two with feasible solution $\mathbf{x}^*$ to find a new optimum $\widetilde{\mathbf{x}}^*$. We henceforth call this procedure the *warm start* of the simplex method. Warm starting the simplex method in phase two might come with the benefit of significantly shortening the number of iterations needed to find $\widetilde{\mathbf{x}}^*$. Conversely, a change in the objective function of the primal problem (LP) corresponds to a change of the dual polyhedron in (DLP) and thus a previously determined optimal dual solution $\boldsymbol{\pi}^*$ is not necessarily dual feasible. Therefore, the dual simplex method generally cannot be warm started given a change in the primal objective function. Note that a change of $\mathbf{b}$ in the primal problem (LP) corresponds to a change of the objective function in (DLP). In such a case, the previously determined solution $\mathbf{x}^*$ does not necessarily stay primal feasible, whereas the dual solution $\boldsymbol{\pi}^*$ remains dual feasible and, therefore, the dual simplex method can be warm started with $\boldsymbol{\pi}^*$. The reoptimization of problem (LP) by means of a warm started simplex method plays an essential role in many implementations of Benders decomposition algorithms (Rahmaniani et al., 2017).

2.2 Mixed-Integer Linear Programming

In *Mixed-Integer Linear Programming* (MILP), additional integer variables are introduced to define a model of the form

$$\text{(MILP)} \qquad \begin{aligned} z^{\text{MILP}} = \min \quad & \mathbf{c}^{\mathrm{T}}\mathbf{x} + \mathbf{d}^{\mathrm{T}}\mathbf{y} && (2.8)\\ \text{s.t.} \quad & \mathbf{A}\mathbf{x} + \mathbf{B}\mathbf{y} \leq \mathbf{b} && (2.9)\\ & \mathbf{x} \in \mathbb{R}^n_{\geq 0} && (2.10)\\ & \mathbf{y} \in \mathbb{Z}^t_{\geq 0} && (2.11) \end{aligned}$$

where $\mathbf{y} \in \mathbb{Z}^t_{\geq 0}$ are non-negative integer-valued decision variables and $\mathbf{B} \in \mathbb{R}^{m \times t}$ is the corresponding coefficient matrix. From a computational perspective, the solution of MILP models is a lot more challenging than the solution of LP models. In fact, solving (MILP) is NP-hard. An important concept about solving (MILP) is the notion of the *convex hull* of the feasible region in (MILP). Let $S^{\text{MILP}} = \{(\mathbf{x}, \mathbf{y}) \in \mathbb{R}^n_{\geq 0} \times \mathbb{Z}^t_{\geq 0} : \mathbf{A}\mathbf{x} + \mathbf{B}\mathbf{y} \leq \mathbf{b}\}$ be the feasible region of (MILP). The convex hull of S^{MILP}, denoted by $\text{conv}(S^{\text{MILP}})$, is the smallest convex set to contain all $(\mathbf{x}, \mathbf{y}) \in S^{\text{MILP}}$. By definition, an optimal solution of (MILP) is attained at a vertex of $\text{conv}(S^{\text{MILP}})$ and is given by the solution of the LP $\min\{\mathbf{c}^{\mathrm{T}}\mathbf{x} + \mathbf{d}^{\mathrm{T}}\mathbf{y} : (\mathbf{x}, \mathbf{y}) \in \text{conv}(S^{\text{MILP}})\}$. However, determining $\text{conv}(S^{\text{MILP}})$ is hard, and therefore, in practice, one tries to approximate the convex hull to solve (MILP) (Wolsey, 2008). Two popular solution approaches that solve (MILP) exactly by iteratively approximating the convex hull are the *branch-and-bound* and *branch-and-cut* algorithms. Both the branch-and-bound and the branch-and-cut algorithm are respectively discussed in the following sections.

2.2.1 Branch-and-Bound

The idea of the branch-and-bound (B&B) algorithm is to divide the original model (MILP) into smaller subproblems while considering the entire solution space of (MILP) through implicit enumeration. To efficiently optimize subproblems, the integrality requirement on variables $\mathbf{y}$ in problem (MILP) is relaxed which leads to an LP model that is called the *LP relaxation* of (MILP). The LP relaxation of (MILP) can efficiently be solved by methods like the simplex method. The optimal objective value of the LP relaxation of (MILP) is a so-called *dual bound* $\underline{z}$ on the optimal objective value of (MILP). In B&B, LP relaxations are respectively solved in nodes of a so-called *B&B tree*. Let $\mathbf{y}'$ be a solution vector to the solved LP relaxation in an arbitrary node. Whenever a solution of a node is fractional for some y_i' of $\mathbf{y}'$, the subproblem is *branched* on variable y_i' by creating two child nodes and adding constraints $y_i \leq \lfloor y_i' \rfloor$ and $y_i \geq \lceil y_i' \rceil$ to the LP relaxations of those child nodes, respectively. If an integer solution is found after solving an LP relaxation in a node of the B&B tree, a *primal bound* $\overline{z}$ on the optimal value of (MILP) is established. After an integer solution is found in a node, the B&B tree is *pruned* after that node, i.e., child nodes are not further explored, since no further improvement in terms of objective function (2.8) could be made. Next to pruning by integrality, the B&B tree may also be pruned by infeasibility and by bound (Wolsey, 2021). The reason for pruning infeasible nodes is that child nodes of an infeasible node cannot be feasible. Nodes are pruned by bound when the best-known primal bound is better than the objective value of a node with a fractional solution since objective values cannot further improve in child nodes. By following the aforementioned rules for the pruning of nodes, the implicit enumeration of all possible integer solutions is ensured. After solving a node of the B&B tree, the relative gap Δ with

$$\Delta = \begin{cases} 0 & \text{if } \underline{z} = \overline{z} = 0 \\ \infty & \text{if } \overline{z} = 0, \underline{z} \neq 0 \\ \frac{|\overline{z} - \underline{z}|}{|\overline{z}|} & \text{otherwise} \end{cases} \tag{2.12}$$

is computed. It is worth noting that various definitions for Δ are available. For this thesis, we adopt the definition used by Gurobi Optimization, LLC (2023) as specified in (2.12). It is worth mentioning that Δ is only defined if a primal bound

$\overline{z}$ is available. The B&B process terminates as soon as Δ falls below a predefined threshold $\Delta^{\min}$. Note that B&B is regarded as a heuristic when terminated with $\Delta^{\min} > 0$ and as an exact algorithm for $\Delta^{\min} = 0$.

2.2.2 Branch-and-Cut

A branch-and-cut (B&C) algorithm is a B&B algorithm, in which so-called *cutting planes* (short *cuts*) are generated throughout the B&B tree when processing nodes (Wolsey, 2021). Cuts are inequalities that are *valid* for (MILP) and are employed to improve the dual bound by cutting off fractional solutions of the LP relaxation at the current node of the B&C tree. Definition 1 provides a formal definition of valid inequalities and cutting planes.

Definition 1 (Valid Inequalities and Cutting Planes). *An inequality* $\mathbf{e}^{\mathrm{T}}\mathbf{x} + \mathbf{f}^{\mathrm{T}}\mathbf{y} \leq e_0$ *is valid for* S^{MILP} *if and only if* $\mathbf{e}^{\mathrm{T}}\mathbf{x} + \mathbf{f}^{\mathrm{T}}\mathbf{y} \leq e_0 \ \forall (\mathbf{x}, \mathbf{y}) \in S^{\mathrm{MILP}}$*. A valid inequality* $\mathbf{e}^{\mathrm{T}}\mathbf{x} + \mathbf{f}^{\mathrm{T}}\mathbf{y} \leq e_0$ *for* S^{MILP} *is a cutting plane if and only if* $\mathbf{e}^{\mathrm{T}}\mathbf{x}^* + \mathbf{f}^{\mathrm{T}}\mathbf{y}^* > e_0$ *for some* $(\mathbf{x}^*, \mathbf{y}^*) \in \mathbb{R}^n_{\geq 0} \times \mathbb{R}^t_{\geq 0} : \mathbf{A}\mathbf{x} + \mathbf{B}\mathbf{y} \leq \mathbf{b}$.

If S^{MILP} and a point $(\mathbf{x}^*, \mathbf{y}^*) \in \mathbb{R}^n_{\geq 0} \times \mathbb{R}^t_{\geq 0} : \mathbf{A}\mathbf{x} + \mathbf{B}\mathbf{y} \leq \mathbf{b}$ are given, the problem of showing that $(\mathbf{x}^*, \mathbf{y}^*) \in \mathrm{conv}(S^{\mathrm{MILP}})$, or to find a valid inequality $\mathbf{e}^{\mathrm{T}}\mathbf{x} + \mathbf{f}^{\mathrm{T}}\mathbf{y} \leq e_0$ cutting off $(\mathbf{x}^*, \mathbf{y}^*)$ from S^{MILP} with $\mathbf{e}^{\mathrm{T}}\mathbf{x}^* + \mathbf{f}^{\mathrm{T}}\mathbf{y}^* > e_0$ is called the *separation problem* (Wolsey, 2008). Separated cuts added to the LP relaxation possibly improve the dual bounds at nodes of the B&C tree which might in turn lead to more pruned nodes and thus smaller B&C trees. Therefore, typically, fewer nodes have to be explored than in B&B, at the expense of the increased processing effort of nodes. Furthermore, B&C implementations often not only encompass the deployment of cuts but other techniques like preprocessing or primal heuristics throughout the B&C process. Therefore, employing B&C instead of B&B involves a philosophical shift. Rather than quickly exploring large parts of the B&B tree through fast reoptimization of LP relaxations, more time and computational effort are allocated to individual nodes with the aim to obtain stronger dual bounds and to potentially enhance primal solutions (Wolsey, 2021). The employment of cuts, preprocessing and primal heuristics in B&C implementations alongside various other options are discussed in Wolsey (2021).

2.3 Benders Decomposition

Depending on the size of (MILP) and other properties like the strength of its corresponding LP relaxation, solving (MILP) within reasonable time frames becomes computationally intractable, even for solvers employing state-of-the-art B&C methods. However, particular problem structures allow for the application of model decomposition techniques that break down the original model (MILP) into multiple other problems that are potentially easier to solve collectively. In this section, we introduce Benders decomposition as one of those model decomposition techniques and explore relevant acceleration techniques for implementing Benders decomposition algorithms.

Benders decomposition can be applied when a model (MILP) has the form

$$
(\mathrm{MILP}^{\mathrm{BD}}) \qquad
\begin{aligned}
z^{\mathrm{MILP}^{\mathrm{BD}}} = \min \quad & \mathbf{c}^{\mathrm{T}}\mathbf{x} + \mathbf{d}^{\mathrm{T}}\mathbf{y} && (2.13)\\
\text{s.t.} \quad & \mathbf{A}\mathbf{x} + \mathbf{B}\mathbf{y} \leq \mathbf{b} && (2.14)\\
& \mathbf{D}\mathbf{y} \leq \mathbf{e} && (2.15)\\
& \mathbf{x} \in \mathbb{R}^{n}_{\geq 0} && (2.16)\\
& \mathbf{y} \in \mathbb{Z}^{t}_{\geq 0}. && (2.17)
\end{aligned}
$$

In this case, such models have a *single-bordered block-diagonal structure*, where constraints $\mathbf{Dy} \leq \mathbf{e}$ correspond to the *first stage* and constraints $\mathbf{Ax} + \mathbf{By} \leq \mathbf{b}$ correspond to the second stage of the problem (Maher, 2021). We call variables $\mathbf{y}$ in ($\mathrm{MILP}^{\mathrm{BD}}$) *linking variables* as they appear in both stages of the problem. Another term for variables $\mathbf{y}$ are *complicating variables* as without the integrality requirement on $\mathbf{y}$, ($\mathrm{MILP}^{\mathrm{BD}}$) would be an LP problem and thus efficiently solvable (Nemhauser and Wolsey, 1988). The single-bordered block-diagonal structure of model (MILP) is depicted in Fig. 2.1. The master block represents the first stage and contains constraints $\mathbf{Dy} \leq \mathbf{e}$. The linking block and blocks $1, \ldots, n^{\mathrm{sp}}$ represent the second stage with constraints $\mathbf{Ax} + \mathbf{By} \leq \mathbf{b}$. The idea of Benders decomposition is based on the idea of fixing variables $\mathbf{y}$ to values $\bar{\mathbf{y}} \in Y = \left\{\mathbf{y} \in \mathbb{Z}^{t}_{\geq 0} : \mathbf{Dy} \leq \mathbf{e}\right\}$ (Nemhauser and Wolsey, 1988). As observable in Fig. 2.1, as soon as variables $\mathbf{y}$ attain fixed values $\bar{\mathbf{y}}$, the problem in the second stage can be further decomposed into blocks $1, \ldots, n^{\mathrm{sp}}$ as blocks are then decoupled. In the following we assume, without loss of generality, that ($\mathrm{MILP}^{\mathrm{BD}}$) has an optimal solution with finite objective value $z^{\mathrm{MILP}^{\mathrm{BD}}}$.

Like Nemhauser and Wolsey (1988), we now assume linking variables $\mathbf{y}$ to be fixed to values $\bar{\mathbf{y}} \in Y$ to reformulate problem ($\mathrm{MILP}^{\mathrm{BD}}$). The resulting LP is the

Benders subproblem (SP) that contains constraints and variables of the second stage. Note that, here, for the sake of simplicity, we only discuss the case of a single Benders SP. However, the problem structure may allow the decomposition with more than one Benders SP, cf. Fig. 2.1. In that case, the concepts described here are directly extendable to several Benders SPs. The resulting Benders SP (SP) is given by

$$z^{\mathrm{SP}}(\bar{\mathbf{y}}) = \min \quad \mathbf{c}^{\mathrm{T}}\mathbf{x} \tag{2.18}$$

$$(\mathrm{SP}) \qquad \text{s.t.} \quad \mathbf{A}\mathbf{x} + \mathbf{B}\bar{\mathbf{y}} \le \mathbf{b} \tag{2.19}$$

$$\mathbf{x} \in \mathbb{R}^{n}_{\ge 0}. \tag{2.20}$$

Since (SP) is part of the original model (MILP^{BD}) and obtained by fixing variables $\mathbf{y}$ to values $\bar{\mathbf{y}} \in Y$, problem (SP) is often called the *inner minimization problem* of (MILP^{BD}). Based on the concept of duality introduced in Sect. 2.1.1, we state the dual Benders SP (DSP) that corresponds to (SP) with

$$z^{\mathrm{DSP}}(\bar{\mathbf{y}}) = \max \quad \boldsymbol{\pi}^{\mathrm{T}}\left(\mathbf{b} - \mathbf{B}\bar{\mathbf{y}}\right) \tag{2.21}$$

$$(\mathrm{DSP}) \qquad \text{s.t.} \quad \boldsymbol{\pi}^{\mathrm{T}}\mathbf{A} \le \mathbf{c}^{\mathrm{T}} \tag{2.22}$$

$$\boldsymbol{\pi} \in \mathbb{R}^{m}_{\le 0}. \tag{2.23}$$

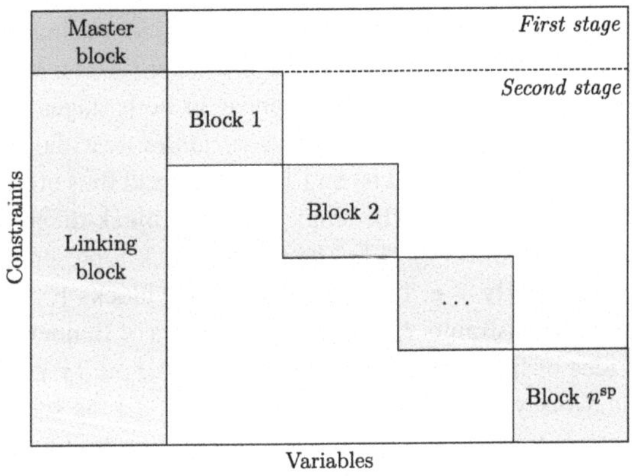

Fig. 2.1 Single-bordered block-diagonal structure of model (MILP^{BD}). The problem is divided into two stages. The master block corresponds to the first stage and contains constraints $\mathbf{D}\mathbf{y} \le \mathbf{e}$. The linking block and blocks $1, \ldots, n^{\mathrm{sp}}$ correspond to the second stages and contain constraints $\mathbf{A}\mathbf{x} + \mathbf{B}\mathbf{y} \le \mathbf{b}$. Variables $\mathbf{y}$ appear in both stages and therefore inherently link the constraints of the two stages

We denote the feasible region of (DSP) with $\Pi = \{\boldsymbol{\pi} \in \mathbb{R}^m_{\leq 0} : \boldsymbol{\pi}^\mathrm{T}\mathbf{A} \leq \mathbf{c}^\mathrm{T}\}$. Note that in (DSP), the feasible region Π does not depend on values of $\bar{\mathbf{y}}$. In the following, we assume that (DSP) is feasible, i.e., $\Pi \neq \emptyset$. Note that $\Pi \neq \emptyset$ is a valid assumption as otherwise, the inner minimization problem (SP) would be either unbounded or infeasible for any value of $\bar{\mathbf{y}}$, contradicting the finite value of $z^{\mathrm{MILP^{BD}}}$. The dual polyhedron Π in (DSP) can be represented by the set of dual extreme rays $\{\boldsymbol{\pi}_r\}_{r\in R}$ and the set of dual extreme points $\{\boldsymbol{\pi}_e\}_{e\in E}$. A dual extreme ray $\boldsymbol{\pi}_r \in \mathbb{R}^m_{\leq 0}$ for $r \in R$ is a direction of unboundedness for Π in (DSP) with $\boldsymbol{\pi}_r^\mathrm{T}(\mathbf{b} - \mathbf{B}\bar{\mathbf{y}}) > 0$ and $\boldsymbol{\pi}_r^\mathrm{T}\mathbf{A} \leq \mathbf{0}^\mathrm{T}$. Note that by Farkas' Lemma, cf. Theorem 2, the existence of dual extreme ray $\boldsymbol{\pi}_r$ exactly corresponds to the infeasibility of (SP). Conversely, an extreme point $\boldsymbol{\pi}_e \in \mathbb{R}^m_{\leq 0}$ for $e \in E$ yields the objective value $\boldsymbol{\pi}_e^\mathrm{T}(\mathbf{b} - \mathbf{B}\bar{\mathbf{y}}) = z^{\mathrm{DLP}}(\bar{\mathbf{y}})$. With sets R and E, we reformulate the original model ($\mathrm{MILP^{BD}}$) to give rise to the Benders master problem (MP) given by

$$
\text{(MP)} \quad
\begin{aligned}
z^{\mathrm{MP}} = \min \quad & \eta + \mathbf{d}^\mathrm{T}\mathbf{y} && && (2.24)\\
\text{s.t.} \quad & \mathbf{D}\mathbf{y} \leq \mathbf{e} && && (2.25)\\
& \boldsymbol{\pi}_r^\mathrm{T}(\mathbf{b} - \mathbf{B}\mathbf{y}) \leq 0 && \forall r \in R && (2.26)\\
& \boldsymbol{\pi}_e^\mathrm{T}(\mathbf{b} - \mathbf{B}\mathbf{y}) \leq \eta && \forall e \in E && (2.27)\\
& \mathbf{y} \in \mathbb{Z}^t_{\geq 0} && && (2.28)\\
& \eta \in \mathbb{R}. && && (2.29)
\end{aligned}
$$

To derive (MP), the inner minimization problem (SP) of the original model ($\mathrm{MILP^{BD}}$) is first substituted by the corresponding dual problem (DSP). Then, the dual polyhedron Π is represented by including inequalities (2.26) and (2.27) with sets R and E into problem (MP). Inequalities (2.26) are so-called *Benders feasibility cuts* as they inhibit solutions of (MP) in variables $\mathbf{y}$ that yield unboundedness of (DSP) and thus infeasibility of (SP) and ($\mathrm{MILP^{BD}}$). Similarly, inequalities (2.27) are so-called *Benders optimality cuts* as they introduce lower bounds on the continuous variable η depending on variables $\mathbf{y}$. Variable η is then used in the objective function (2.24). Henceforth, we say that Benders feasibility cuts (2.26) and Benders optimality cuts (2.27) together make up the set of *Benders cuts* of problem (MP).

As shown, model (MP) solves the original model ($\mathrm{MILP^{BD}}$) equivalently. However, the cardinalities $|R|$ and $|E|$ are generally very large, making solving (MP) intractable from a computational standpoint. Therefore, Benders (1962) proposes to solve the problem in an iterate manner by only considering subsets $R' \subseteq R$ and $E' \subseteq E$ in each iteration step by relaxing problem (MP). In the following section, we introduce this iterative solution approach in its classical form as proposed by

Benders (1962). Furthermore, in most implementations using Benders decomposition, a combination of different acceleration techniques is used. Rahmaniani et al. (2017) introduce a taxonomy for the categorization of such acceleration techniques into categories *Decomposition Strategy*, *Solution Procedure*, *Solution Generation* and *Cut Generation*. In the following, we explore relevant concepts related to the last three categories mentioned (presented in Sects. 2.3.2, 2.3.3, and 2.3.4, respectively). It is important to note that while we utilize the taxonomy from Rahmaniani et al. (2017) for categorization, we also discuss additional concepts not covered in their work. For a comprehensive overview of Benders decomposition applications and its acceleration techniques, we refer to the literature review of Rahmaniani et al. (2017).

2.3.1 Classical Implementation

As previously mentioned, solving model (MP) is computationally intractable. Therefore, in the classical implementation of the Benders decomposition algorithm, the model is solved for subsets $R' \subseteq R$ and $E' \subseteq E$. This leads to the *relaxed Benders master problem* (RMP) defined by

$$\text{(RMP)} \quad \begin{aligned} z^{\text{RMP}} = \min \quad & \eta + \mathbf{d}^{\mathrm{T}}\mathbf{y} && && (2.30)\\ \text{s.t.} \quad & \mathbf{D}\mathbf{y} \leq \mathbf{e} && && (2.31)\\ & \boldsymbol{\pi}_r^{\mathrm{T}}(\mathbf{b} - \mathbf{B}\mathbf{y}) \leq 0 && \forall r \in R' && (2.32)\\ & \boldsymbol{\pi}_e^{\mathrm{T}}(\mathbf{b} - \mathbf{B}\mathbf{y}) \leq \eta && \forall e \in E' && (2.33)\\ & \mathbf{y} \in \mathbb{Z}^t_{\geq 0} && && (2.34)\\ & \eta \in \mathbb{R}. && && (2.35) \end{aligned}$$

The classical implementation of the Benders decomposition algorithm based on the work of Benders (1962) is shown in Algorithm 1 in which we still assume $z^{\text{MILP}^{\text{BD}}}$ being finite. In the case of a minimization problem like the one in (RMP), the classical Benders decomposition algorithm starts by setting the upper bound UB and lower bound LB to ∞ and $-\infty$, respectively. Subsequently, (RMP) is solved with $R' = E' = \emptyset$ which then yields an optimal solution $(\bar{\mathbf{y}}, \bar{\eta})$. With solution $(\bar{\mathbf{y}}, \bar{\eta})$, LB can be updated since problem (RMP) is a relaxation of (MP) for which $z^{\text{MP}} = z^{\text{MILP}^{\text{BD}}}$. Afterward, the dual problem (DSP) is solved to optimality by fixing linking variables in (DLP) to $\bar{\mathbf{y}}$. If (DSP) is unbounded, i.e., $z^{\text{DSP}}(\bar{\mathbf{z}}) = \infty$, we add r, corresponding to dual extreme ray $\boldsymbol{\pi}_r$, to the current index set of Benders feasibility cuts R'. Otherwise, if $z^{\mathrm{DSP}}(\bar{\mathbf{z}}) > \bar{\eta}$, we add e, corresponding to the dual extreme point $\boldsymbol{\pi}_e$, to the current index set of Benders optimality cuts E'. Furthermore, note that UB is updated in the case the original model (RMP) is feasible for $(\bar{\mathbf{y}}, z^{\mathrm{DSP}}(\bar{\mathbf{y}}))$. The algorithm then iteratively solves (RMP) and (DSP) with the previously described procedure until $UB - LB \leq \varepsilon$. Benders (1962) shows that the algorithm terminates within a finite number of iterations based on the cardinalities $|R|$ and $|E|$ in (MP) being finite. Note that Algorithm 1 always terminates when $z^{\mathrm{DSP}}(\bar{\mathbf{y}}) = \bar{\eta}$ as then $LB = UB$.

Algorithm 1 Classical Benders decomposition algorithm based on Benders (1962)

$UB \leftarrow \infty$
$LB \leftarrow -\infty$
$R' \leftarrow \emptyset$
$E' \leftarrow \emptyset$
while $UB - LB > \varepsilon$ **do**
  solve (RMP) with sets R' and E' to optimality and obtain $(\bar{\mathbf{y}}, \bar{\eta})$
  $LB \leftarrow \mathbf{d}^{\mathrm{T}}\bar{\mathbf{y}} + \bar{\eta}$
  with $(\bar{\mathbf{y}}, \eta)$, solve (DSP) to optimality and obtain $z^{\mathrm{DSP}}(\bar{\mathbf{y}})$
  if $z^{\mathrm{DLP}}(\bar{\mathbf{y}}) = \infty$ **then**
    retrieve dual extreme ray $\boldsymbol{\pi}_r$ from solution of (DSP)
    $R' \leftarrow R' \cup \{r\}$
  else
    if $z^{\mathrm{DLP}}(\bar{\mathbf{y}}) > \bar{\eta}$ **then**
      retrieve dual extreme point $\boldsymbol{\pi}_e$ from solution of (DSP)
      $E' \leftarrow E' \cup \{e\}$
    $UB \leftarrow \min\{UB, \mathbf{d}^{\mathrm{T}}\bar{\mathbf{y}} + z^{\mathrm{DSP}}(\bar{\mathbf{y}})\}$

Importantly, in the classical implementation of the Benders decomposition algorithm presented in Algorithm 1 problem (DSP) is solved to compute Benders feasibility and optimality cuts. However, as long as dual extreme rays $\boldsymbol{\pi}_r$ or dual extreme points $\boldsymbol{\pi}_e$ can be computed, one can also solve the primal problem (SP) (Maher, 2021). As introduced in Sect. 2.1.2, the simplex method can be used to optimize problem (SP) and yield dual solutions $\boldsymbol{\pi}_r$ and $\boldsymbol{\pi}_e$. For clarity, we henceforth only write that (DSP) is optimized to compute dual extreme rays $\boldsymbol{\pi}_r$ or dual extreme points $\boldsymbol{\pi}_e$.

2.3.2 Solution Procedure

Rahmaniani et al. (2017) identify that the iterative solution procedure of the previously described classical Benders decomposition algorithm is a major computational bottleneck. While (RMP) is commonly repeatedly solved by B&B or B&C, (DSP) is typically optimized by the simplex method. In this section, we show several approaches on how to adapt the classical Benders decomposition algorithm with the goal of making better use of the B&B or B&C process as well as the simplex method.

Branch-and-Cut Approach

In the classical implementation of Benders decomposition, problem (RMP) is repeatedly solved to optimality. This reoptimization creates a lot of computational overhead as for each optimization, a new B&B tree is created (Maher, 2021). Also, an optimal solution of the original model ($MILP^{BD}$) might even be found in earlier iterations but is not identified as an optimal solution as it is discarded in the B&B process of the solver since necessary Benders cuts are still missing in (RMP) to prove optimality (Maher, 2021). For this reason, a relatively novel approach is to integrate the Benders decomposition algorithm in the B&C process of solvers which leads to a substantial reduction in computation time in numerous studies (Rahmaniani et al., 2017). This approach is henceforth called *branch-and-Benders-cut* (B&BC). Generally, there are two ways to apply B&BC. Firstly, some MILP solvers have a general purpose B&BC algorithm implemented that automatically decomposes a given model and applies B&BC while also providing the user some options to adjust the decomposition or parameters of the algorithm. Examples of commercial and non-commercial MILP solvers to employ B&BC in an automatic Benders decomposition framework are CPLEX and SCIP (Bonami et al., 2020; Maher, 2021). Secondly, many solvers offer different callback functions to interact with the solver while performing B&C. This enables the user to implement their own Benders cut separation routines during B&C, effectively making it a B&BC algorithm. Maher (2021) presents a non-exhaustive list of commercial and non-commercial MILP and constraint integer programming solvers that fulfill the requirements for implementing a user-made Benders cut separation routine in B&BC. Note that Benders cuts need to be valid for the original model ($MILP^{BD}$) but may cut off integer solutions of (RMP). Therefore, Benders cuts stand in contrast to user cuts that must not cut off any integer solution of (RMP) and are solely employed to improve the dual bound of nodes in B&BC.

While problem (RMP) is only solved a single time in B&BC, problem (DSP) is typically optimized more often than in the classical implementation (Rahmaniani et al., 2017). Therefore, using B&BC instead of the classical approach promises reductions in solution time if the optimization time of (DSP) is short relative to (RMP) (Rahmaniani et al., 2017). Furthermore, Benders cuts can be separated at different stages in the B&BC process. For every integer solution of (RMP) during the B&BC process problem (DSP) needs to be optimized to separate Benders cuts as otherwise solutions of (RMP) would be regarded as feasible for the original model (MILP^{BD}) even if they were not. Beyond separating Benders cuts for integer solutions of (RMP), it is also possible to separate Benders cuts for fractional solutions of (RMP) and thus theoretically at every node of the tree. However, doing so often leads to a rapidly increasing size of (RMP) and might make the optimization of its LP relaxations computationally intractable.

Reoptimization of Benders Subproblems

Depending on the Benders SP and the chosen optimization method, it might be possible to significantly reduce optimization time by using previously optimal solutions for the reoptimization of the Benders SP. In the case of Benders SPs like (DSP), the simplex method is typically used for optimization (Rahmaniani et al., 2017). During the Benders decomposition algorithm, variables $\mathbf{y}$ in Benders SPs are fixed to a solution vector $\bar{\mathbf{y}}$ discovered in (RMP). Fixing variables $\mathbf{y}$ only affects the objective function of (DSP) while the dual polyhedron Π remains unchanged. Therefore, as explained in Sect. 2.1.2, the dual simplex method can be warm started in phase two with a previously feasible solution. Often, a new optimum can be found with only a few simplex iterations, thereby significantly reducing the time needed for reoptimization.

Parallelizing Optimization of Benders Subproblems

As previously depicted in Fig. 2.1, problem (SP) can sometimes be decomposed into several independent Benders SPs. In this case, a parallel optimization of Benders SPs might lead to a significant reduction in computation time required for the optimization of Benders SPs in comparison with the optimization of a single Benders SP. While further decomposing the single Benders SP into multiple Benders SPs bears the potential to reduce optimization time, Chermakani (2015) observes that it might be better to aggregate some of the Benders SPs when the number of Benders SPs is significantly larger than the number of available processors.

If for an integer solution of (RMP), a single separated Benders cut is violated by that solution, other Benders cuts pertaining to different Benders SPs theoretically do not need to be separated and added to the sets of Benders cuts in (RMP) as the current solution is already invalidated. Therefore, (RMP) can be reoptimized as soon as a violated Benders cut has been added. This bears the potential for trade-offs as the invalidation of an integer solution of (RMP) by fewer Benders cuts might reduce the computation time and number of Benders cuts needed. However, separating more Benders cuts could further strengthen the LP relaxation of (RMP), potentially leading to faster convergence.

2.3.3 Solution Generation

The number of iterations in the classical Benders decomposition algorithm or the size of the tree in B&BC depends on the quality of solutions found for linking variables $\mathbf{y}$ as they are used to generate Benders cuts and bounds (Rahmaniani et al., 2017). In this section, we introduce acceleration techniques that focus on finding solutions for (RMP) more quickly or that aim at improving the formulation of (RMP).

Two-phase Approach

Due to model complexity, finding an optimal solution of (RMP) for the classical Benders decomposition algorithm or integer solutions of (RMP) for B&BC is typically hard. For this reason, McDaniel and Devine (1977) propose to apply a two-phase approach for the classical Benders decomposition algorithm. In the first phase, problem (RMP) is relaxed by dropping the integrality requirement of variables $\mathbf{y}$. Then, the relaxed version of (RMP) is used to quickly separate Benders cuts to tighten the formulation. Then, in the second phase, the integrality requirement on integer variables $\bar{\mathbf{y}}$ is reintroduced and the classical Benders decomposition algorithm is used as usual and therefore yields an optimal solution of the original model.

In the context of B&BC, the two-phase approach of McDaniel and Devine (1977) can be applied by separating Benders cuts for fractional solutions at the root node of the search tree with the aim to reduce the resulting tree size (Rahmaniani et al., 2017).

Valid Inequalities

By decomposing the original model (MILP^{BD}) into problems (RMP) and (DSP), some information on the original solution structure of linking variables $\mathbf{y}$ in (MILP^{BD}) might not be maintained in (RMP). Some of this information is then iteratively retained by adding Benders cuts that are outer approximations of the solution space of (DSP) and therefore imply further restrictions on linking variables $\mathbf{y}$. This lack of information might lead to slow convergence of Benders decomposition algorithms as the relevant information is only slowly retained by Benders cut separation. Depending on the problem structure, it is possible to strengthen (RMP) by adding valid inequalities to impose some restrictions $\mathbf{y}$ initially projected out to (DSP), e.g., by excluding some solutions $\bar{\mathbf{y}}$ that lead to unboundedness of (DSP).

Note that, analogously to Benders cuts, such valid inequalities only need to be valid for the original model (MILP^{BD}) and not for (RMP). Depending on the specific Benders decomposition, it is, therefore, possible to include valid inequalities that make problem (DSP) feasible for all $\bar{\mathbf{y}} \in Y$. Such valid inequalities therefore avoid the separation of Benders feasibility cuts and possibly lead to an acceleration of the overall optimization process. Different references where valid inequalities are employed to strengthen the LP relaxation of (RMP) can be found in Rahmaniani et al. (2017).

Multicuts

If problem (DSP) is further decomposable (cf. Fig. 2.1), Benders cuts can be generated for each Benders SP. In this case, Benders cuts are often collectively called *multicuts* (Rahmaniani et al., 2017). Birge and Louveaux (1988) show that the use of multicuts in a classical Benders decomposition algorithm significantly reduces the maximum theoretical number of iterations when compared to the generation of a single, aggregated Benders cut. Furthermore, Birge and Louveaux (1988) observe that the number of iterations until termination of the classical Benders decomposition algorithm is also significantly reduced in practice. However, simultaneously, using multicuts lead to larger sizes of (RMP) which might increase the needed time in following iterations. While several implementations show that the separation of multicuts may lead to faster convergence (Contreras et al., 2011; Jenabi et al., 2015), this trade-off remains to be considered.

In case n^{sp} multicuts are separated in each iteration of the algorithm, a vector of variables $\eta \in \mathbb{R}^{n^{sp}}$ is introduced to (RMP) instead of a single continuous variable $\eta \in \mathbb{R}$. Furthermore, note that even though for parallelization of Benders SPs the decomposability of problem (DSP) is also required, the parallelization of Benders SPs is not necessary for the generation of multicuts, i.e., depending on the aggregation of Benders SPs, more than one Benders cut might be separated per Benders SP.

2.3.4 Cut Generation

The number of iterations in the classical Benders decomposition algorithm or the tree size in B&BC is closely related to the strength of Benders cuts (Rahmaniani et al., 2017). In this section, we introduce the issue of dual degenerate solutions of Benders SPs and briefly introduce the in-out method as a procedure to stabilize Benders cut separation at the root node and strengthen Benders cuts during B&BC.

Dual Degeneracy and Pareto-Optimal Benders Cuts

In the presence dual degeneracy in (DSP), multiple optimal solutions of (DSP) exist that yield different Benders cuts of potentially different strength. Magnanti and Wong (1981) show that the selection of the Benders cut from different optimal degenerate dual solutions can significantly impact the convergence of the classical Benders decomposition algorithm. In their work, Magnanti and Wong (1981) introduce a framework for assessing the strength of Benders optimality cuts, introducing the concepts of *cut domination* and *Pareto-optimality* as given in Definition 2.

Definition 2 (Cut domination and Pareto-optimality (Magnanti and Wong, 1981)). *A Benders optimality cut* (2.27) *corresponding to* $\boldsymbol{\pi}_1 \in \Pi$ *dominates that corresponding to* $\boldsymbol{\pi}_2 \in \Pi$ *if*

$$\boldsymbol{\pi}_1^{\mathrm{T}}(\mathbf{b} - \mathbf{B}\bar{\mathbf{y}}) \geq \boldsymbol{\pi}_2^{\mathrm{T}}(\mathbf{b} - \mathbf{B}\bar{\mathbf{y}}) \quad \forall \bar{\mathbf{y}} \in Y \tag{2.36}$$

with strict inequality for at least one point $\bar{\mathbf{y}} \in Y$*. In this case, we also say that* $\boldsymbol{\pi}_1$ *dominates* $\boldsymbol{\pi}_2$*. A cut is Pareto-optimal if it is not dominated by any other cut. Similarly, we say that the point* $\boldsymbol{\pi}$ *that corresponds to the cut is Pareto-optimal.*

Further, Magnanti and Wong (1981) introduce the *Magnanti-Wong problem* that requires a *core point* $\mathbf{y}^c$ of Y. A core point $\mathbf{y}^c$ is a point in the relative interior of the convex hull of Y. The authors show that solving the Magnanti-Wong problem allows for the separation of a Pareto-optimal Benders optimality cut.

Extending Algorithm 1, Magnanti and Wong (1981) develop an algorithm in which the Magnanti-Wong problem is solved every time after solving (DSP) to generate Pareto-optimal Benders optimality cuts. Even though optimizing the Magnanti-Wong problem in every iteration increases computational effort, the authors show in a case study that their method may lead to faster convergence by lowering the number of total iterations needed.

Several authors identified disadvantages of the method of Magnanti and Wong (1981). On the one hand, solving an additional subproblem increases time and computational effort needed for the separation of Benders cuts. On the other hand, the Magnanti-Wong problem is prone to be numerically unstable (Papadakos, 2008; Sá et al., 2013). Lastly, identifying a core point $\mathbf{y}^c$ might be difficult, depending on the problem under investigation (Rahmaniani et al., 2017). Several authors adapt the method of Magnanti and Wong (1981) to overcome the aforementioned issues. Papadakos (2008) introduces an adaption of the Magnanti-Wong problem that does not require solving (DSP), thus circumventing the need to solve an additional subproblem. Furthermore, Papadakos (2008) shows that different points called

Magnanti-Wong points can be used to separate Pareto-optimal cuts and that any convex combination of Magnanti-Wong points is a Magnanti-Wong point. Sherali and Lunday (2013) show that a perturbation of the right-hand-side of (SP) is sufficient to separate maximal nondominated Benders optimality cuts, also avoiding solving the additional subproblem as in Magnanti and Wong (1981). All the aforementioned studies rely on the existence of a core point $\mathbf{y}^c$ or a Magnanti-Wong point. Furthermore, some authors introduce methods to approximate core points (Santoso et al., 2005; Papadakos, 2008). However, those approximations rely on the classical Benders algorithm shown in Algorithm 1.

In-Out Method

Integer programs, large-scale linear programs, and convex problems are mostly solved by adding inequalities violated by a determined optimum of the current LP relaxation (Ben-Ameur and Neto, 2007). As problem sizes become larger, methods that use this type of cutting plane mechanisms, especially Benders decomposition in its classical application, are known to behave poorly as many inequalities are separated and the lower bound only improves very slowly (Ben-Ameur and Neto, 2007). For the classical Benders decomposition algorithm, especially large initial steps in early iterations as well as excessive oscillation when the algorithm approaches the optimum are identified as reasons for this phenomenon (Rahmaniani et al., 2017). For this reason, many authors have focused on stabilizing the solution process of (RMP), e.g, by means of regularized decomposition, trust-regions, and level decomposition strategies (Rahmaniani et al., 2017).

A method that can be used to stabilize the cut process is the in-out method where a stabilization point $\mathbf{y}^0$ is used to determine a separation point $\mathbf{y}^{\text{sep}}$ that lies between the current LP relaxation solution $\bar{\mathbf{y}}$ and the stabilization point $\mathbf{y}^0$. The stabilization point $\mathbf{y}^0$ is iteratively updated by a convex combination of the previous value for $\mathbf{y}^0$ and the new LP relaxation solution $\bar{\mathbf{y}}$. Fischetti et al. (2017) implement an in-out method based on the work of Ben-Ameur and Neto (2007) and employ it to separate Benders cuts before solving the problem with B&BC. After every five iterations, they remove Benders cuts with slack to prevent the Benders RMP from growing excessively in size. Furthermore, Fischetti et al. (2017) apply a distortion vector $\boldsymbol{\delta}$ to the separation point if the best known dual bound does not change after five iterations, and after five more iterations without improvement in the dual bound they revert to the cut separation scheme of Kelley, i.e., the separation of Benders cuts with $\bar{\mathbf{y}}$ (Kelley, 1960), until no improvement is made again for five iterations. The authors find that the in-out method significantly reduces the size of the resulting B&BC tree and accelerates the optimization process. Bonami et al. (2020) implement a similar in-out method for the stabilization of the initial cut loop into the commercial solver CPLEX.

Conversely, Maher (2021) implements an in-out method similar to Fischetti et al. (2017) into the solver SCIP, that, however, is used for the strengthening of Benders cuts separated at different nodes of the B&BC tree. The author finds that this method compares favorably with the regular separation of Benders cuts and poses an effective alternative to other Benders cut strengthening or selection techniques.

Problem and Model Development

3

In this chapter, we introduce the problem of the optimization of transformation pathways for a residential building stock and develop MILP models for this problem. First, in Sect. 3.1, we lay out the inherent multi-layer structure of the problem and then give a formal problem description based on the identified layers. Subsequently, we introduce further definitions and enlist parameters and sets used throughout our MILP models in Sect. 3.2. We then present a MILP model for the introduced problem in Sect. 3.3.

3.1 Problem Description

In this section, we present a problem description for the optimization of transformation pathways for a residential building stock. We first introduce the multi-level structure of the optimization problem and define the goals of the optimization. Consecutively, we formally describe the problem on each of the identified levels.

3.1.1 Multi-level Problem Structure and Optimization Goals

The problem of determining optimal transformation pathways of the residential building stock can be seen as one level of an interdependent multi-level optimization problem. This multi-level optimization problem consists of three logical levels that also exhibit different time resolutions with respect to decision-making and physical processes. The structure of this multi-level optimization problem is depicted in Fig. 3.1. On the first level, *resource allocation at building stock level*, restricted resources need to be allocated to different buildings of the building stock over a given time horizon. Those resources constitute building energy system devices and

R. Delorme, *Optimization of Transformation Pathways for the Residential Building Sector*, BestMasters, https://doi.org/10.1007/978-3-658-47466-9_3

workforce and are respectively restricted every year. Depending on allocated devices and workforce at the first level, so-called *transformation pathways* of individual buildings are determined on the second level. Transformation pathways comprise *building efficiency pathways* and *building energy system pathways*. Building efficiency pathways define states and their transitions of building efficiency along a time horizon with a yearly resolution while building energy system pathways define states and their transitions of the building energy system design for every year of the same time horizon. Lastly, on the third level, *building operations* of building energy systems are optimized for every considered year based on daily and hourly operation. Building operations thus depend on the chosen building energy system design and building efficiency in every year and work on a time axis with daily and hourly resolution.

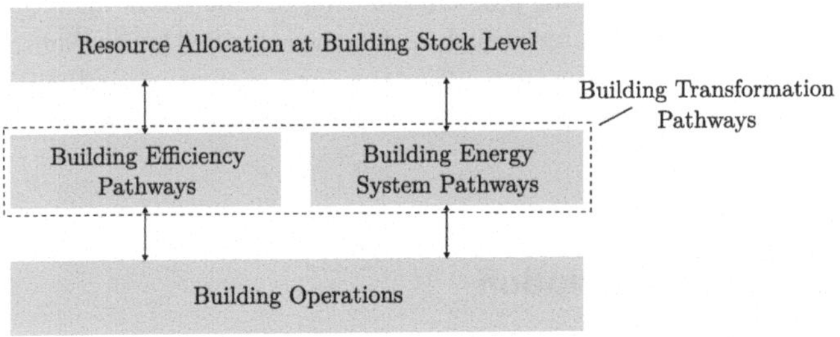

Fig. 3.1 Multi-level structure of the optimization problem. The problem consists of three interdependent levels *resource allocation at building stock level*, *building transformation pathways*, and *building operations*. Building transformation pathways comprise both *building efficiency pathways* and *building energy system pathways*

The inherent interdependence of the three problem levels is schematically depicted in Fig. 3.2. Allocated resources in the form of devices and workforce in every year determine transitions in building efficiency or building energy system design. Here, workforce is required for the installation of new devices, the realization of efficiency measures, or the conductance of energy consultancies in buildings. Building operations of buildings depend on the current building efficiency and building energy system design of a given year. Moreover, both building efficiency and building energy system pathways are temporally interconnected over the years. For building efficiency pathways, the temporal interconnection exists in the form of transitions between building efficiency states through newly realized efficiency measures. Analogously, the temporal interconnection for building energy system pathways exists in the form of transitions between states of the building energy system by means of newly installed devices or devices that reached their lifetime. Therefore, to optimally solve the described problem, all three levels must be considered simultaneously within the optimization process.

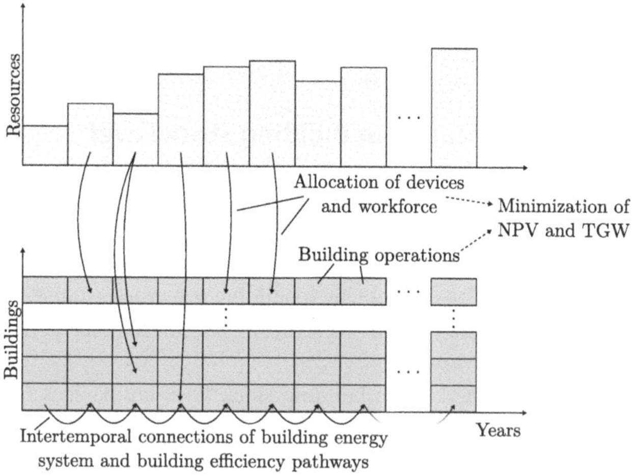

Fig. 3.2 Problem of determining optimal allocations of devices and workforce onto buildings of the residential building stock over a given time horizon. The allocation of devices and workforce determines transitions in building efficiency and building energy system pathways. Depending on the states of building efficiency and the building energy system within respective pathways, building operations are optimized for every building and every year of the considered time horizon. Therefore, the problem is intertemporally connected by means of building efficiency and building energy system pathways

Similar to the work of Richarz et al. (2022), the optimization goal is to investigate trade-offs between economical and ecological aspects of transformation pathways for residential building stock. Therefore, we choose to optimize convex combinations of the net present value (NPV) as well as the total global warming (TGW) of the entire building stock. The NPV is the aggregated value of all capital flows, discounted to the present point in time with a yearly discount rate of ρ. The NPV comprises investment costs, costs for operations and maintenance (O&M), and energy costs. We define the TGW to be an impact indicator for climate change and let the TGW aggregate direct and indirect CO_2-equivalent (CO_2eq) emissions with their respective global warming potentials (GWPs) (Heijungs, 2014). We do not assume a discount rate for CO_2eq emissions, mainly to avoid an ethical debate on how to conduct such a discounting[1]. As depicted in Fig. 3.2, the NPV is impacted by the installation of devices or the realization of efficiency measures as well as by energy costs for building operations. Likewise, the TGW is impacted by the realization of efficiency measures through embodied emissions and direct CO_2eq emissions using building operations.

Following the presented problem structure, we now formally describe the optimization problem for each element of the problem depicted in 3.1.

3.1.2 Resource Allocation at Building Stock Level

For each building $b \in B = \{0, \ldots, n^{\text{buil}} - 1\}$, there are different devices $u \in U$ and efficiency measures $m \in M$ that can be respectively installed or realized. The resource allocation problem at the building stock level consists in deciding in which year $i \in I = \{0, \ldots, n^{\text{years}} - 1\}$ and which building $b \in B$ to install device $u \in U$ or realize efficiency measure $m \in M$. Henceforth, we call the time period covered by years $i \in I$ the *optimization time horizon.*

Resources to be allocated at the building stock level are devices for building energy systems as well as workforce that is needed for the installation of devices, the realization of efficiency measures, and the conductance of energy consultancies in buildings. Because of restrictions in mobility, we assume that workforce is restricted on a regional level. To model regional dependencies on workforce availability, we partition the set B into sets $B_r \subseteq B$ by introducing geographical regions $r \in R = \{0, \ldots, n^{\text{reg}} - 1\}$. For each set B_r, we consider regional supply constraints to restrict the availability of workforce. For each geographical region $r \in R$, there are workforce supplies of $q_{r,i}^{\text{work,dev}}$ and $q_{r,i}^{\text{work,mea}}$ work hours for the installation of devices in buildings or realization of efficiency measures in year $i \in I$, respectively. The installation of device $u \in U$ is assumed to require d_u^{dev} working hours and is therefore building-independent. Conversely, necessary working hours $d_{b,m}^{\text{mea}}$ for the realization of efficiency measure $m \in M$ depend on building $b \in B$ because of significantly different building geometries and conditions. Restrictions in the number of devices to be installed are assumed to hold for the entire building stock since transportation of building energy system devices is not assumed to constitute a significant problem. For a given year $i \in I$, there are in total $q_{u,i}^{\text{dev}}$ devices of device $u \in U$ that can be installed for the entire building stock. Whenever a device $u \in U$ is installed or an efficiency measure $m \in M$ is realized in building $b \in B$ and year

[1] We refer the interested reader to the work of Lueddeckens et al. (2022) for a recent discussion on discounting in life-cycle assessment.

$i \in I$, an energy consultancy must take place in the same year for that building. We assume the effort of an energy consultancy to be building-dependent with d_b^{encon} work hours for $b \in B$. The supply of total work hours for energy consultancies in region $r \in R$ and year $i \in I$ is given by $q_{r,i}^{\text{work,encon}}$.

3.1.3 Building Efficiency Pathways

Realizing an efficiency measure $m \in M$ in a building might induce a transition in building efficiency and therefore energy demands. For each building $b \in B$, we consider efficiency measures for the roof, windows, and walls that can be realized (Richarz et al., 2022). For this purpose, we introduce the set of efficiency measure classes $C^{\text{mea}} = \{\text{roof,wall,win}\}$. Furthermore, sets $S_c = \{0, \ldots, n_c^{\text{stage}} - 1\}$ represent sets of stages of each efficiency measure class $c \in C^{\text{mea}}$. Different stages $s \in S_c$ relate to different characteristics, e.g., efficiency level, costs, embodied emissions, etc., of efficiency measures pertaining to the same class $c \in C^{\text{mea}}$. We then define sets M_c of efficiency measures pertaining to class c as

$$M_c = \{c\} \times S_c \qquad \forall c \in C^{\text{mea}} \tag{3.1}$$

and the set of all efficiency measures M as

$$M = \bigcup_{c \in C^{\text{mea}}} M_c. \tag{3.2}$$

In each building $b \in B$ and year $i \in I$, exactly one efficiency measure $m \in M_c$ is contained in every $c \in C^{\text{mea}}$. If a new efficiency measure m is realized, the previously realized efficiency measure of efficiency measure class $c \in C^{\text{mea}} : m \in M_c$ is replaced by the new efficiency measure for the current and following years. Based on the notion of the sets M_c, we introduce the set of building efficiencies E, defined as

$$E = \prod_{c \in C^{\text{mea}}} M_c, \tag{3.3}$$

i.e., the set E is the Cartesian product of sets M_c for $c \in C^{\text{mea}}$. Building efficiencies $e \in E$, therefore, contain exactly one measure $m \in M_c$ for every $c \in C^{\text{mea}}$. For every year $i \in I$ and building $b \in B$, exactly one efficiency $e \in E$ needs to be taken. The building efficiency of building b at the beginning of the optimization time horizon is given by e_b^{exeff}. Moreover, for existing efficiency measures contained in e_b^{exeff}, the remaining lifetime of efficiency measures is denoted by $l_{b,m}^{\text{exmea}} > 0$. If

$l_{b,m}^{\text{exmea}} < n^{\text{years}} - 1$ for some $m \in M$, an efficiency measure of the same efficiency measure class $c \in C^{\text{mea}} : m \in M_c$ needs to be realized in at least one year $i \in \{0, \ldots, l_{b,m}^{\text{exmea}} - 1\}$.

The realization of efficiency measures $m \in M$ in building $b \in B$ and year $i \in I$ incurs total investment costs of $c_{b,m,i}^{\text{mea}}$. The lifetime of efficiency measure $m \in M$ is l_m^{mea} years, including the year of realization and we assume $l_m^{\text{mea}} \geq n^{\text{years}}$ so that the residual value at the end of the optimization time horizon needs to be accounted for. Furthermore, the realization of efficiency measure $m \in M$ in building $b \in B$ incurs total embodied CO_2eq emissions of $f_{b,m}^{\text{mea}}$.

3.1.4 Building Energy System Pathways

At the individual building level, there are several design decisions to be made concerning the building energy system. The set of devices U that may be installed in every building $b \in B$ defines the so-called *superstructure* of the building energy system. This superstructure is depicted in Fig. 3.3. For the production of thermal energy, we consider heat pumps with air as energy source (HP^{air}), heat pumps with geothermal energy as an energy source (HP^{geo}), boilers working with gas (BOI^{gas}), oil (BOI^{oil}) and pellets (BOI^{pel}), combined heat and power (CHP) with gas as an energy source, electrical heaters (EH), and solar thermal collectors (STC). Devices comprising the set for the generation of electrical energy are solar photovoltaic modules (PV) and CHP. Devices STC and PV to be installed on the roof of buildings define the set of roof devices $U^{\text{roof}} = \{\text{STC,PV}\}$. Both electrical and thermal energy may be respectively stored using a battery (BAT) or thermal energy storage (TES). With devices BAT and TES, we define the set of storages $U^{\text{sto}} = \{\text{BAT,TES}\}$. Some buildings may have an existing connection to district heating with which they can satisfy their thermal energy demands. Furthermore, every building is connected to the electricity grid with which it may exchange electrical energy.

Based on the superstructure depicted in Fig. 3.3, we define the set of final energy demands $F = \{\text{el,th}\}$ where el and th denote electrical and thermal energy, respectively. Moreover, we define the set of energy carriers that can potentially be imported by buildings with the set $G^{\text{imp}} = \{\text{el,gas,oil,pel,th}\}$ where pel denotes pellets. The set of importable energy carriers $G_b^{\text{imp}} \subseteq G^{\text{imp}}$ differs depending on whether building $b \in B$ is connected to district heating. In case building b is connected to district heating, it is $G_b^{\text{imp}} = \{\text{el,gas,oil,pel,th}\}$. Conversely, if $b \in B$ is not connected to district heating, we define $G_b^{\text{imp}} = \{\text{el,gas,oil,pel}\}$. Analogously, we define the set of energy carriers that can be exported by every building $b \in B$ to

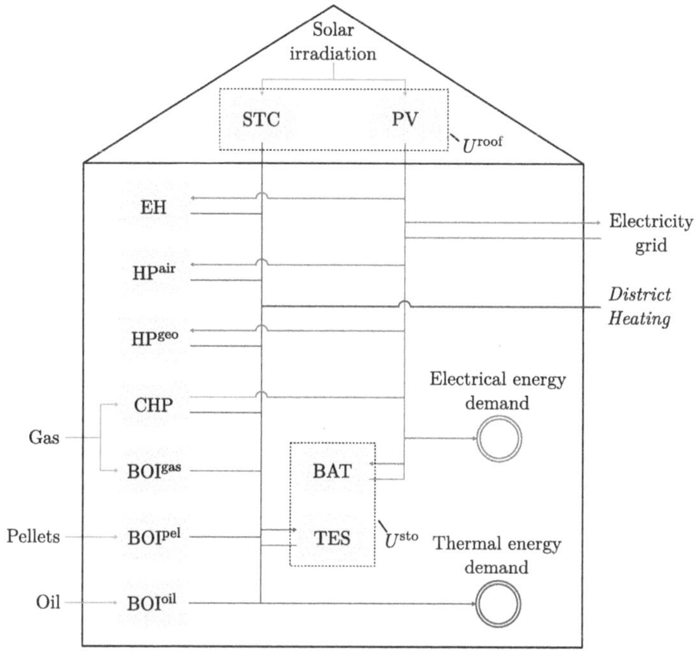

Fig. 3.3 Considered superstructure of buildings $b \in B$. In each building, both thermal and electrical energy demands need to be satisfied. Thermal energy may be generated by components HP^{air}, HP^{geo}, CHP, BOI^{gas}, BOI^{oil}, BOI^{pel}, EH or STC. Furthermore, some buildings may have an existing connection to district heating to satisfy thermal energy demands. Electrical energy is produced by components CHP and PV or is exchanged with the electricity grid. Components BAT and TES can be used to temporarily store electrical or thermal energy, respectively

be $G^{\text{exp}} = \{\text{el}\}$. Furthermore, let sets $H_u^{\text{in}} \subseteq G^{\text{imp}} \cup F$ and $H_u^{\text{out}} \subseteq G^{\text{imp}} \cup F$ for all $u \in U$ be the sets of energy carriers flowing in and out of device u, respectively. The concrete definition of sets H_u^{in} and H_u^{out} for devices $u \in U$ is provided in Table 3.1.

Every device $u \in U$ is assumed to have a lifetime of l_u^{dev} that includes the year of installation. When installed in year $i \in I$, the size of the device may be chosen up to a value of $o_{b,u}^{\text{dev}}$ and can be used in years $j \in I_{u,i}^{\text{dev}} = \{i, \ldots, \min\{i + l_{u,i}^{\text{dev}}, n^{\text{years}}\} - 1\}$. The physical property described by the size of the device depends on the type of the device. Here, we make the following distinction:

Table 3.1 Sets of inflowing and outflowing energy carriers H_u^{in} and H_u^{out} for devices $u \in U$

$u \in U$	H_u^{in}	H_u^{out}
BAT	{el}	{el}
BOI^{gas}	{gas}	{th}
BOI^{oil}	{oil}	{th}
BOI^{pel}	{pel}	{th}
CHP	{gas}	{el,th}
EH	{el}	{th}
HP^{air}	{el}	{th}
HP^{geo}	{el}	{th}
STC	∅	{th}
TES	{th}	{th}
PV	∅	{el}

- For items $u \in U^{\text{roof}}$, the size of the device refers to the covered roof area by device u.
- For items $u \in U^{\text{sto}}$, the size of the device refers to the total amount of internal energy stored by device u.
- For items $u \in U \setminus (U^{\text{sto}} \cup U^{\text{roof}})$, the size of the device refers to the output energy flow rate of u.

Furthermore, we introduce the set of device classes C^{dev} that contains subsets $C' \subseteq U$ that comprise technology-wise similar devices. We define the set of device classes C^{dev} with

$$C^{\text{dev}} = \Big\{\{\text{BAT}\}, \{\text{BOI}^{\text{gas}}, \text{BOI}^{\text{oil}}, \text{BOI}^{\text{pel}}\}, \{\text{CHP}\}, \{\text{EH}\}, \{\text{HP}^{\text{air}}, \text{HP}^{\text{geo}}\}, \{\text{STC}\}, \{\text{TES}\}, \{\text{PV}\}\Big\}. \tag{3.4}$$

Following the work of Schütz (2018), we assume that for devices $u \in U \setminus U^{\text{roof}}$, only a single device of each technology class $C' \in C^{\text{dev}}$ may be available in each year $i \in I$. Therefore, in case a device $u \in U \setminus U^{\text{roof}}$ is installed in year $i \in I$ in building $b \in B$ and a second device $v \in U \setminus U^{\text{roof}}$ is installed in an arbitrary year $j \in I_{u,i}^{\text{dev}} \setminus \{i\}$, device u with its corresponding size is replaced and cannot be used anymore in years $k \in I_{u,i}^{\text{dev}} \cap I_{v,j}^{\text{dev}}$. We alter from this assumption for devices $u \in U^{\text{roof}}$ that are assumed to remain usable for all years $i \in I_{u,i}^{\text{dev}}$ due to their inherent modular extendibility. However, installed devices $u \in U^{\text{roof}}$ must not cover more area than given by the building-specific area a_b^{roof} in any year $i \in I$.

In every building $b \in B : \text{th} \notin G_b^{\text{imp}}$ peak demands $p_{b,p}^{\text{peak}}$ for $p \in P = \{\text{dhw}, \text{th}\}$ need to be covered in every year by components $U_p^{\text{heat}} \subseteq U$. The peak heat demand for DHW $p_{b,\text{dhw}}^{\text{peak}}$ is covered by devices $U_{\text{dhw}}^{\text{heat}} = \{\text{BOI}^{\text{gas}}, \text{BOI}^{\text{oil}}, \text{BOI}^{\text{pel}}, \text{CHP}, \text{EH}\}$ that can heat up DHW to a constant temperature of t^{dhw}. Like Richarz et al. (2022), we assume a constant cold water temperature of t^{cw} and a constant flow water temperature of t^{fl}. The peak heat demand for DHW is then given by $p_{b,\text{dhw}}^{\text{peak}} = \tilde{t}^{\text{dhw}} p_b^{\text{dhw,max}}$ where $\tilde{t}^{\text{dhw}} = \frac{t^{\text{dhw}} - t^{\text{fl}}}{t^{\text{dhw}} - t^{\text{cw}}}$ is the fraction of required energy flow rate to heat water by devices $u \in U_{\text{dhw}}^{\text{heat}}$ from t^{cw} to t^{dhw} under the assumption of a constant specific heat capacity. The peak heat demand for thermal energy $p_{b,\text{th}}^{\text{heat}}$, given by maximum required thermal energy flow for space heating and DHW, can be covered by devices $U_{\text{th}}^{\text{heat}} = \{\text{BOI}^{\text{gas}}, \text{BOI}^{\text{oil}}, \text{BOI}^{\text{pel}}, \text{EH}, \text{CHP}, \text{HP}^{\text{air}}, \text{HP}^{\text{geo}}\}$ that are assumed to be reliably able to serve peak thermal energy demands. Buildings with district heating $b \in B : \text{th} \in G_b^{\text{imp}}$ are assumed to be able to cover peak heat demands $p_{b,p}^{\text{peak}}$ for $p \in P$ reliably as we assume no maximum heat flow rate from district heating and the heat to be at a temperature level high enough for DHW to reach t^{dhw}. For electrical energy demands, we do not define such a requirement for any building due to the possibility to cover peak demands with imports of electrical energy from the grid. We assume that for devices $u \in U \setminus U^{\text{roof}} : \text{th} \in H_u^{\text{out}}$ it holds that $o_{b,u}^{\text{dev}} \geq \max_{p \in P}\{p_{b,p}^{\text{peak}}\}$, i.e., if device u is available, can be used to cover $p \in P$, and sized adequately, it may cover the peak demands $p_{b,p}^{\text{peak}}$.

For each building $b \in B$, we consider an existing building energy system design comprising different devices $u \in U$. Henceforth, *existing* devices denote devices that exist at the beginning of the considered optimization time horizon. All existing devices have a remaining lifetime of $l_{b,u}^{\text{exdev}} > 0$ and it is $l_{b,u}^{\text{exdev}} = 0$ if device $u \in U$ does not exist in building $b \in B$. Existing devices $u \in U \setminus U^{\text{roof}}$ may be used in years $i \in I_{b,u}^{\text{exdev}} = \{0, \ldots, \min\{l_{b,u}^{\text{exdev}}, n^{\text{years}}\} - 1\}$ as long as they are not replaced by another device of the same technology class. An existing device $u \in U \setminus U^{\text{roof}}$ has an available unit size of $o_{b,u}^{\text{exdev}}$, while devices $u \in U^{\text{roof}}$ cover a roof area of $a_{b,u}^{\text{roofdev}}$ until they reach their respective lifetime in building b.

Similar to Richarz et al. (2022), the installation of device $u \in U$ in year $i \in I$ is assumed to incur fixed investment costs of $c_{u,i}^{\text{dev,fix}}$ and size-dependent variable investment costs of $c_{u,i}^{\text{dev,var}}$ per unit size. When the lifetime of a device is reached or it is replaced by another device of the same technology class, we do not assume any additional removal costs. For costs of O&M, we consider a device-specific factor

c_u^{om} that represents the factor of investment costs spent for O&M of devices. Costs $c_u^{\text{om}} c_{u,i}^{\text{dev,fix}}$ of O&M related to fixed investment costs for devices occur in each year a device is used and relate to investment costs in the year of installation $i \in I$. Costs of O&M for device $u \in U$ in year $i \in I$ related to variable investment costs, however, are proportional to the maximum used device size of the respective year $i \in I$ with a factor of c_u^{om}. For an existing device $u \in U$ in $b \in B$ with $l_{u,b}^{\text{exdev}} > 0$, costs of O&M are assumed to be given by $c_u^{\text{om}} c_{b,u}^{\text{exdev}}$ where $c_{b,u}^{\text{exdev}}$ are the total investment costs of device $u \in U$ in buildings $b \in B$.

3.1.5 Building Operations

For building operations, we consider a set of typical days D where days $d \in D$ are assumed to adequately represent a number of $w_{b,d}^{\text{day}}$ days of the original year for building $b \in B$. Each typical day consists of a set of time steps T and we assume the building energy system to operate on a resolution given by time step length Δ^{ts}. Most commonly, in the context of energy systems optimization, the period length and time step length are set to days with hourly periods, mainly due to the availability of data at this time resolution (Teichgraeber and Brandt, 2022). Moreover, relatively high time resolutions are important to appropriately represent the integration potential of renewable energies in energy systems (Kotzur et al., 2021). Energy demand for $f \in F$ must be satisfied in every time step by available devices $u \in U : f \in H_u^{\text{out}}$ or the electricity grid for $f = \text{el}$. Therefore, the decision to be made in this stage lies in deciding on how to exactly operate available devices $u \in U$ that are available in $b \in B$ and $i \in I$ such that energy demands $f \in F$ are satisfied in all time steps. The availability of devices therefore couples building energy system pathways to building operations. The energy conversion of device $u \in U$ from energy input to the energy output of $f \in H_u^{\text{out}}$ is assumed with a capacity- and power independent device efficiency of $\eta_{b,u,f,d,\tau}$, e.g., for device CHP, we assume two device efficiencies $\eta_{b,\text{CHP},\text{el},d,\tau}$ and $\eta_{b,\text{CHP},\text{th},d,\tau}$ that denote the share of energy outflow rate per unit of energy inflow rate for $f = \text{el}$ and $f = \text{th}$, respectively.

For the distribution system of $f \in F$ inside the building, we assume quasi-stationarity, i.e., the change in stored electrical or thermal energy in distribution systems is zero between time steps $\tau - 1$ and τ. Like Schütz (2018) and Kotzur (2018), we assume for devices $u \in U^{\text{sto}}$ that a share of el_u of the stored energy is lost in every time step. Furthermore, the depth-of-discharge (DOD), i.e., the share of storage capacity until which storage $u \in U^{\text{sto}}$ can be discharged, is given with dod_u. During charging and discharging of storage $u \in U^{\text{sto}}$, we assume constant charging and discharging efficiencies of η_u^{ch} and η_u^{dch}, respectively. The minimum time to fully discharge the accessible share of a full energy storage $u \in U^{\text{sto}}$ is given by Δ_u^{sto} and is used to restrict both charging and discharging rates of the energy storage under the assumption of constant charging and discharging rates. We assume that for every typical day $d \in D$, the amount of stored energy in storage $u \in U^{\text{sto}}$ before the first and after the last time step needs to be equally high to close the daily energy balance (Kotzur, 2018). For devices $u \in U^{\text{roof}}$, the input energy flow rate is given by the net area on the roof times the normal solar irradiation $g_{b,d,\tau}$ for building $b \in B$, typical day $d \in D$ and time step $\tau \in T$.

For a given building $b \in B$, the demand of electrical energy is independent of building efficiency $e \in E$ and given by $p_{b,d,\tau}^{\text{el}}$ for typical day $d \in D$ and time step $\tau \in T$. The demand for thermal energy comprises both the demand due to domestic hot water (DHW) $p_{b,d,\tau}^{\text{dhw}}$ as well as the demand for space heating $p_{b,e,d,\tau}^{\text{heat}}$ for building efficiency $e \in E$ of the given year $i \in I$. For the peak demands $p_{b,p}^{\text{peak}}$ for $p \in P$, we assume that $p_{b,\text{th}}^{\text{peak}} \geq \max_{e\in E, d\in D, \tau\in T}\{p_{b,d,\tau}^{\text{dhw}} + p_{b,e,d,\tau}^{\text{heat}}\}$ as well as $p_{b,\text{dhw}}^{\text{peak}} \geq \tilde{t}^{\text{dhw}} \max_{d\in D, \tau\in T}\{p_{b,d,\tau}^{\text{dhw}}\}$ for all $b \in B$. The demand for space heating $p_{b,e,d,\tau}^{\text{heat}}$ depends on the building efficiency $e \in E$ that is attained in a given year $i \in I$ for building $b \in B$. This element couples buildings operations to building efficiency pathways. Lastly, and analogously to Richarz et al. (2022), to avoid the forming of legionella, in every time step $\tau \in T$, it needs to be ensured that a share of $\tilde{t}^{\text{dhw}}$ of the necessary energy flow rate for DHW demand $p_{b,d,\tau}^{\text{dhw}}$ is covered by devices $u \in U_{\text{dhw}}^{\text{heat}}$ to heat DHW up to a temperature of t^{dhw}.

The energy costs and direct CO_2eq emissions for the import of one energy unit of $g \in G^{\text{imp}}$ in year $i \in I$ are given with $c_{g,i}^{\text{en}}$ and $f_{g,i}^{\text{en}}$. Conversely, one exported energy unit of $g \in G^{\text{exp}}$ entails revenues $s_{g,i}^{\text{en}}$. Furthermore, the export of one energy unit of $g \in G^{\text{exp}}$ is assumed to avoid direct CO_2eq emissions of $f_{g,i}^{\text{en}}$.

3.2 Definitions and Notations

In this section, we first introduce new sets to facilitate the MILP modeling of the described problem. Then, we provide a complete overview of the sets and parameters used for the MILP models in this chapter.

For devices $u \in U$ and years $i \in I$, we introduce sets $I_{u,i}^{\text{avail}} = \{\max\{0, i - l_{u,i}^{\text{dev}} + 1\}, \ldots, i\}$ to denote years $j \in I_{u,i}^{\text{avail}}$ in which device u needs to be installed to be available in year i. Furthermore, for all $b \in B$ and $m \in M$, we define sets $I_{b,m}^{\text{exmea}} = \{0, \ldots, \min\{l_{b,m}^{\text{exmea}}, n^{\text{years}}\} - 1\}$ that denote the set of years $i \in I_{b,m}^{\text{exmea}}$ in which an existing efficiency measure m in building b can be used.

The sets and parameters are listed alphabetically in Table 3.2 and Table 3.3, respectively. For readability of our MILP models, we denote all parameters in lower case letters while we capitalize all decision variables. Furthermore, we use subscripts solely for indexation and superscripts solely for descriptions.

Table 3.2 Sets used for MILP model formulations

Set	Description
B	Set of buildings
C^{dev}	Set of device classes
C^{mea}	Set of efficiency measure classes
D	Set of typical days
E	Set of building efficiencies
F	Set of final energy carriers
G^{exp}	Set of exportable energy carriers
G^{imp}	Set of all importable energy carriers
G_b^{imp}	Set of importable energy carriers of building b
H_u^{in}	Set of energy carriers flowing in device u
H_u^{out}	Set of energy carriers flowing out of device u
I	Set of years in optimization time horizon
$I_{u,i}^{\text{avail}}$	Set of years in year i in which device u must be installed to be available
$I_{u,i}^{\text{dev}}$	Set of years in which device u is available if installed in year i
$I_{b,u}^{\text{exdev}}$	Set of years in which existing device u is available in building b
$I_{b,m}^{\text{exmea}}$	Set of years in which existing efficiency measure m can be used in building b
M	Set of all efficiency measures
M_c	Set of efficiency measures pertaining to efficiency measure class c
P	Set of peak demands
R	Set of geographical regions
S_c	Set of stages for efficiency measure class c
T	Set of time steps
U	Set of all devices
U_p^{heat}	Set of devices to be used to cover peak demand p
U^{roof}	Set of devices to be installed on the roof of buildings
U^{sto}	Set of storages

Table 3.3 Parameters used for model formulations, their sets, and definitions

Symbol	Sets	Description
$a_{b,i}^{\text{avail}}$	$b \in B, i \in I$	Available roof area in building b and year i
a_b^{roof}	$b \in B$	Roof area in building b
$a_{b,u}^{\text{roofdev}}$	$b \in B, u \in U^{\text{roof}}$	Roof area covered by existing device u in building b
$c_{g,i}^{\text{en}}$	$g \in G^{\text{imp}}, i \in I$	Specific costs of imported energy carrier g in year i
$c_{u,i}^{\text{dev,fix}}$	$u \in U, i \in I$	Fixed investment costs of device u in year i
$c_{u,i}^{\text{dev,var}}$	$u \in U, i \in I$	Variable investment costs of device u in year i for one unit size
$c_{b,u}^{\text{exdev}}$	$b \in B, u \in U$	Total investment costs of existing device u in building b
$c_{b,m,i}^{\text{mea}}$	$b \in B, m \in M,$ $i \in I$	Total investment costs of efficiency measure m in building b in year i
d_u^{dev}	$u \in U$	Working hours for installation of device u
d_b^{encon}	$b \in B$	Working hours for conductance of energy consultancy of building b
$d_{b,m}^{\text{mea}}$	$b \in B, m \in M$	Working hours for realization of efficiency measure m in building b
dod_u	$u \in U^{\text{sto}}$	Depth of charge of storage u
e_b^{exeff}	$b \in B$	Initial building efficiency of building b
el_u	$u \in U$	Relative energy loss of energy storage u per time step Δ^{ts}
$f_{g,i}^{\text{en}}$	$g \in G^{\text{imp}} \cup G^{\text{exp}},$ $i \in I$	Direct or avoided CO_2eq emissions of energy carrier g in year i for one energy unit
$f_{b,m}^{\text{mea}}$	$b \in B, m \in M$	Embodied CO_2eq emissions of efficiency measure m in building b
$g_{b,d,\tau}$	$b \in B, d \in D,$ $\tau \in T$	Normal solar irradiation for building b on day d at time step τ
$h_{u,i}^{\text{dev}}$	$u \in U, i \in I$	Auxiliary parameter for calculation of the residual value of investment costs of device u in year i
$h_{m,i}^{\text{mea}}$	$m \in M, i \in I$	Auxiliary parameter for calculation of the residual value of investment costs of efficiency measure m in year i
l_u^{dev}	$u \in U$	Life time of device u in years
$l_{b,u}^{\text{exdev}}$	$b \in B, u \in U$	Remaining lifetime for existing device u in building b
$l_{b,m}^{\text{exmea}}$	$b \in B, m \in M$	Remaining lifetime of efficiency measure m in building b

(Continued)

Table 3.3 (Continued)

Symbol	Sets	Description
l_m^{mea}	$m \in M$	Life time of efficiency measure m in years
n^{buil}		Number of considered buildings $\|B\|$ for optimization
n^{reg}		Number of considered regions $\|R\|$
n_c^{stage}	$c \in C^{mea}$	Number of considered stages for efficiency measure class c
n^{years}		Number of considered years $\|I\|$ for optimization
$p_{b,d,\tau}^{dhw}$	$b \in B, d \in D, \tau \in T$	Thermal energy demand for DHW of building b on day d at time step τ
$p_{b,d,\tau}^{el}$	$b \in B, d \in D, \tau \in T$	Electrical energy demand of building b on day d at time step τ
$p_{b,e,d,\tau}^{heat}$	$b \in B, e \in E, d \in D, \tau \in T$	Thermal energy demand for space heating of building b on day d at time step τ with building efficiency e
$p_{b,p}^{peak}$	$b \in B, p \in P$	Peak energy demand p in building b
$o_{b,u}^{dev}$	$b \in B, u \in U$	Maximum device size of device u in building b
$o_{b,u}^{exdev}$	$b \in B, u \in U$	Installed device size of existing device u in building b
$q_{u,i}^{dev}$	$u \in U, i \in I$	Market supply for device u in year i
$q_{r,i}^{work,dev}$	$r \in R, i \in I$	Working hours supply for device installations in region r in year i
$q_{r,i}^{work,encon}$	$r \in R, i \in I$	Working hours supply for energy consultancies in region r in year i
$q_{r,i}^{work,mea}$	$r \in R, i \in I$	Working hours supply for realization of efficiency measures in region r in year i
$s_{g,i}^{en}$	$g \in G^{exp}, i \in I$	Price of sold energy carrier g in year i
t^{cw}		Cold water temperature
t^{dhw}		Drinking water temperature
$\tilde{t}^{dhw}$		Fraction of energy flow rate needed to heat water from t^{fl} to t^{dhw}
t^{fl}		Flow water temperature of low temperature heating devices
w_b^{buil}	$b \in B$	Number of buildings in buildings stock represented by building b
$w_{b,d}^{day}$	$b \in B, d \in D$	Number of days in year represented by typical day d for building b

(Continued)

Table 3.3 (Continued)

Symbol	Sets	Description
$z^{\text{dev}}_{u,v}$	$u, v \in U$	Parameter equal to one if and only if devices u and v are both in the same device class $C' \in C^{\text{dev}}$
$z^{\text{mea}}_{m,e}$	$m \in M, e \in E$	Parameter equal to one if and only if efficiency measure m is contained in building efficiency e
Δ^{ts}		Duration of time step
Δ^{sto}_{u}	$u \in U^{\text{sto}}$	Minimum time to fully discharge energy storage u
$\eta_{b,u,f,d,\tau}$	$b \in B$, $u \in U \setminus U^{\text{sto}}$, $f \in H^{\text{out}}_{u}, d \in D$, $\tau \in T$	Efficiency of device u in building b for energy output f on day d at time step τ
η^{ch}_{u}	$u \in U^{\text{sto}}$	Charging efficiency of storage u
η^{dch}_{u}	$u \in U^{\text{sto}}$	Discharging efficiency of storage u
ρ		Discount rate for capital flows

3.3 Model Formulation

In this section, we introduce a MILP model for the introduced problem. For the development of the MILP model, we will use the same multi-level structure of the problem presented in Sect. 3.1 that structures the problem into the allocation of resources at the building stock level, building efficiency pathways, building energy system pathways, and building operations. To facilitate the understanding for the reader, we choose to first present constraints pertaining to the aforementioned elements of the problem and then introduce the objective function and constraints used for the calculation of costs and emissions used in the objective function. We introduce variables at their first appearance in the MILP model and use them henceforth in the following parts of the model. The MILP model lays the foundation for the development of model decompositions in Chap. 4 to be used later within a Benders decomposition algorithm.

3.3.1 Resource Allocation at Building Stock Level

$$\sum_{b \in B_r} \sum_{u \in U} w_b^{\text{buil}} d_u^{\text{dev}} X_{b,u,i}^{\text{dev}} \leq q_{r,i}^{\text{work,dev}} \qquad \forall r \in R, i \in I \tag{3.5}$$

$$\sum_{b \in B_r} \sum_{m \in M} w_b^{\text{buil}} d_{b,m}^{\text{mea}} X_{b,m,i}^{\text{mea}} \leq q_{r,i}^{\text{work,mea}} \qquad \forall r \in R, i \in I \tag{3.6}$$

$$\sum_{b \in B} w_b^{\text{buil}} X_{b,u,i}^{\text{dev}} \leq q_{u,i}^{\text{dev}} \qquad \forall u \in U, i \in I \tag{3.7}$$

$$X_{b,u,i}^{\text{dev}} \leq Y_{b,i}^{\text{encon}} \qquad \forall b \in B, u \in U, i \in I \tag{3.8}$$

$$X_{b,m,i}^{\text{mea}} \leq Y_{b,i}^{\text{encon}} \qquad \forall b \in B, m \in M, i \in I \tag{3.9}$$

$$\sum_{b \in B_r} w_b^{\text{buil}} d_b^{\text{encon}} Y_{b,i}^{\text{encon}} \leq q_{r,i}^{\text{work,encon}} \qquad \forall r \in R, i \in I \tag{3.10}$$

$$\sum_{u \in C'} X_{b,u,i}^{\text{dev}} \leq 1 \qquad \forall b \in B, C' \in C^{\text{dev}} : |C'| \geq 2, i \in I \tag{3.11}$$

$$\sum_{\hat{m} \in M_c} \sum_{i \in I_{b,m}^{\text{exmea}}} X_{b,\hat{m},i}^{\text{mea}} \geq 1 \qquad \forall b \in B, c \in C^{\text{mea}}, m \in M_c : 0 < l_{b,m}^{\text{exmea}} \leq n^{\text{years}} - 1 \tag{3.12}$$

$$X_{b,u,i}^{\text{dev}} \in \{0, 1\} \qquad \forall b \in B, u \in U, i \in I \tag{3.13}$$

$$X_{b,m,i}^{\text{mea}} \in \{0, 1\} \qquad \forall b \in B, m \in M, i \in I \tag{3.14}$$

$$Y_{b,i}^{\text{encon}} \in [0, 1] \qquad \forall b \in B, i \in I \tag{3.15}$$

Constraints (3.5) are knapsack constraints to ensure that workforce supply for region $r \in R$ for the installation of devices is not exceeded in year $i \in I$. In constraints (3.5), we use binary variables $X_{b,u,i}^{\text{dev}}$ that take the value one if and only if device $u \in U$ is installed in building $b \in B$ in year $i \in I$. Similarly, we use knapsack constraints (3.6) to ensure that workforce supply in region $r \in R$ and $i \in I$ is not exceeded for the realization of efficiency measures. Here, we use binary variables $X_{b,m,i}^{\text{mea}}$ that take the value one if and only if efficiency measure $m \in M$ is realized in building $b \in B$ and year $i \in I$. The number of devices of the same device type $u \in U$ to be installed in the same year is restricted through constraints (3.7) that limit the number of installed devices per device type $u \in U$ and year $i \in I$.

Energy consultancies are required in the same year a device is installed or an efficiency measure is realized. We use non-negative continuous variables $Y_{b,i}^{\text{encon}}$ that take value one if device $u \in U$ is installed or efficiency measure $m \in M$ is realized in building $b \in B$ and year $i \in I$ using linking constraints (3.8) and (3.9).

Constraints (3.10) then ensure for every region $r \in R$ and year $i \in I$ that the maximum number of energy consultancies is not exceeded. Next, constraints (3.11) enforce that in each building b and year i, only a single device per device class $C' \in C^{\text{dev}}$ is installed. Lastly, constraints (3.12) ensure that for existing efficiency measures $m \in M$ of building $b \in B$ with a lifetime ending within the optimization time horizon, i.e., $0 < l^{\text{exmea}}_{b,m} \leq n^{\text{years}} - 1$, an efficiency measure $\hat{m} \in M_c$ pertaining to the same efficiency class $c \in C^{\text{mea}} : m \in M_c$ needs to be realized in at least one year $i \in I^{\text{exmea}}_{b,m}$. Since we assume that $l^{\text{mea}}_m \geq n^{\text{years}}$, all efficiency measures realized in a year $i \in I$ do not necessarily need to be replaced within $j \in I : j > i$, which is why we do not enforce the realization of efficiency measures for efficiency measure classes with efficiency measures that are realized within the optimization time horizon. Note that we, in deviation from the structure of the problem definition, include constraints (3.11) and (3.12) already here to facilitate the decomposition of the model later.

3.3.2 Building Efficiency Pathways

$$\sum_{e \in E} \sum_{k=i}^{j} z^{\text{eff}}_{m,e} Y^{\text{eff}}_{b,e,k} \geq (j - i + 1)\Big(X^{\text{mea}}_{b,m,i} - \sum_{\substack{\hat{m} \in M: \\ \hat{m} \neq m}} \sum_{k=i+1}^{j} z^{\text{mea}}_{m,\hat{m}} X^{\text{mea}}_{b,\hat{m},k}\Big) \qquad \forall b \in B, m \in M, i, j \in I : j \geq i \tag{3.16}$$

$$(i + 1) \sum_{j=0}^{i} X^{\text{mea}}_{b,m,j} \geq \sum_{e \in E} \sum_{j=0}^{i} z^{\text{eff}}_{m,e} Y^{\text{eff}}_{b,e,j} \qquad \forall b \in B, m \in M : z^{\text{eff}}_{m,e^{\text{exeff}}_b} = 0, i \in I \tag{3.17}$$

$$\sum_{e \in E} Y^{\text{eff}}_{b,e,i} = 1 \qquad \forall b \in B, i \in I. \tag{3.18}$$

$$Y^{\text{eff}}_{b,e,i} \in \mathbb{R}_{\geq 0} \qquad \forall b \in B, e \in E, i \in I \tag{3.19}$$

For the modeling of building efficiency pathways, we introduce non-negative continuous variables $Y^{\text{eff}}_{b,e,i}$ that take the value one if and only if building efficiency $e \in E$ is taken in building $b \in B$ and year $i \in I$. We further introduce parameters $z^{\text{eff}}_{m,e}$ that are equal to one if and only if efficiency measure m is part of building efficiency e and zero otherwise and as well as parameters

$$z^{\text{mea}}_{m,\hat{m}} = \begin{cases} 1 & \text{if } \exists c \in C^{\text{mea}} : m, \hat{m} \in M_c \\ 0 & \text{otherwise} \end{cases} \qquad \forall m, \hat{m} \in M. \tag{3.20}$$

Constraints (3.16) ensure correct switching of building efficiency states $Y^{\text{eff}}_{b,e,i}$ in dependence of installed efficiency measures. If a measure $m \in M$ is installed in year $i \in I$, constraints (3.16) ensure that a corresponding building efficiency $e \in E : z^{\text{eff}}_{m,e} = 1$ must be taken in the same year and all subsequent years for as long as no other efficiency measure $\hat{m} \in M : z^{\text{mea}}_{m,\hat{m}} = 1$ is realized.

Next, constraints (3.17) ensure that efficiency states $Y^{\text{eff}}_{b,e,i}$ remain at zero for as long as no corresponding efficiency measures $m \in M : z^{\text{mea}}_{m,e^{\text{exeff}}_b} = 0$ is installed and thus ensures correct switching from the initial building efficiency e^{exeff}_b. Constraints (3.18) ensure that exactly one efficiency state is taken in each year. Together, constraints (3.16)–(3.18) implicitly ensure that variables $Y^{\text{eff}}_{b,e,i}$ only take binary values even if defined as non-negative continuous variables.

3.3.3 Building Energy System Pathways

$$\dot{E}^{\text{dev}}_{b,u,i,j} \le o^{\text{dev}}_{b,u} X^{\text{dev}}_{b,u,i} \qquad \forall b \in B, u \in U, i \in I, j \in I^{\text{dev}}_{u,i} \tag{3.21}$$

$$\dot{E}^{\text{dev}}_{b,u,i,j} \le o^{\text{dev}}_{b,u}\Big(1 - \sum_{v \in U} z^{\text{dev}}_{u,v} X^{\text{dev}}_{b,v,j}\Big) \qquad \forall b \in B, u \in U \setminus U^{\text{roof}}, i \in I, j \in I^{\text{dev}}_{u,i} \setminus \{i\} \tag{3.22}$$

$$\dot{E}^{\text{dev}}_{b,u,i,j} \le \dot{E}^{\text{dev}}_{b,u,i,j-1} \qquad \forall b \in B, u \in U, i \in I, j \in I^{\text{dev}}_{u,i} \setminus \{i\} \tag{3.23}$$

$$\dot{E}^{\text{exdev}}_{b,u,i} \le a^{\text{roofdev}}_{b,u} \qquad \forall b \in B, u \in U^{\text{roof}}, i \in I^{\text{exdev}}_{b,u} \tag{3.24}$$

$$\dot{E}^{\text{exdev}}_{b,u,i} \le o^{\text{exdev}}_{b,u}\Big(1 - \sum_{v \in U} z^{\text{dev}}_{u,v} X^{\text{dev}}_{b,v,j}\Big) \qquad \forall b \in B, u \in U \setminus U^{\text{roof}}, i \in I^{\text{exdev}}_{b,u}, j \in \{0, \ldots, i\} \tag{3.25}$$

$$\dot{E}^{\text{avail}}_{b,u,i} = \sum_{j \in I^{\text{avail}}_{u,i}} \dot{E}^{\text{dev}}_{b,u,j,i} + \dot{E}^{\text{exdev}}_{b,u,i} \qquad \forall b \in B, u \in U, i \in I^{\text{exdev}}_{b,u} \tag{3.26}$$

$$\dot{E}^{\text{avail}}_{b,u,i} = \sum_{j \in I^{\text{avail}}_{u,i}} \dot{E}^{\text{dev}}_{b,u,j,i} \qquad \forall b \in B, u \in U, i \in I \setminus I^{\text{exdev}}_{b,u} \tag{3.27}$$

$$p^{\text{peak}}_{b,p} \le \sum_{u \in U^{\text{heat}}_p} \dot{E}^{\text{avail}}_{b,u,i} \qquad \forall b \in B : \text{th} \notin G^{\text{imp}}_b, p \in P, i \in I \tag{3.28}$$

$$a^{\text{avail}}_{b,i} \ge \sum_{u \in U^{\text{roof}}} \sum_{j \in I^{\text{avail}}_{u,i}} \dot{E}^{\text{dev}}_{b,u,j,j} \qquad \forall b \in B, i \in I \tag{3.29}$$

$$\dot{E}^{\text{dev}}_{b,u,i,j} \in \mathbb{R}_{\ge 0} \qquad \forall b \in B, u \in U, i \in I, j \in I^{\text{dev}}_{u,i} \tag{3.30}$$

$$\dot{E}^{\text{exdev}}_{b,u,i} \in \mathbb{R}_{\ge 0} \qquad \forall b \in B, u \in U, i \in I^{\text{exdev}}_{b,u} \tag{3.31}$$

$$\dot{E}^{\text{avail}}_{b,u,i} \in \mathbb{R}_{\ge 0} \qquad \forall b \in B, u \in U, i \in I \tag{3.32}$$

For the calculation of the available device size in a given year, we introduce non-negative continuous variables $\dot{E}^{\text{dev}}_{b,u,i,j}$ that represent the usable device size in building $b \in B$ for a device $u \in U$ installed in year $i \in I$ and used in year $j \in I^{\text{dev}}_{u,i}$. Constraints (3.21) ensure that the device size $\dot{E}^{\text{dev}}_{b,u,i,j}$ can be greater than zero for years $j \in I^{\text{dev}}_{u,i}$ only if device $u \in U$ is installed in year $i \in I$ and restrict its size to $o^{\text{dev}}_{b,u}$. Furthermore, we model that for devices $u \in U \setminus U^{\text{roof}}$, only the last installed device of a technology class may be used. For this, constraints (3.22) force the value of $\dot{E}^{\text{dev}}_{b,u,i,j}$ to zero in case another device of that technology class is installed in year $j \in I^{\text{dev}}_{u,i} \setminus \{i\}$. For constraints (3.23), we define parameters $z^{\text{dev}}_{u,v}$ by

$$z^{\text{dev}}_{u,v} = \begin{cases} 1 & \text{if } \exists C' \in C^{\text{dev}} : u, v \in C' \\ 0 & \text{otherwise} \end{cases} \qquad \forall u, v \in U. \tag{3.33}$$

Constraints (3.23) then enforce that values for $\dot{E}^{\text{dev}}_{b,u,i,j}$ stay at zero for subsequent years $j' \in I^{\text{dev}}_{u,i} : j' > j$.

If in building $b \in B$ device $u \in U \setminus U^{\text{roof}}$ exists before the optimization time horizon, i.e., $l^{\text{exdev}}_{b,u} > 0$ and $o^{\text{exdev}}_{b,u} > 0$, this device is available for as long as no other device of the same technology class is installed. We use non-negative continuous variables $\dot{E}^{\text{exdev}}_{b,u,i}$ that represent the available device size of existing device $u \in U$ in building $b \in B$ and year $i \in I^{\text{exdev}}_{b,u}$. Constraints (3.24) ensure that existing devices $u \in U^{\text{roof}}$ can be used in years $i \in I^{\text{exdev}}_{b,u}$ up to the installed area of $a^{\text{roofdev}}_{b,u}$. Similarly to constraints (3.22), we use constraints (3.25) for devices $u \in U \setminus U^{\text{roof}}$ to ensure that variables $\dot{E}^{\text{exdev}}_{b,u,i}$ take value zero for the year a device of the same technology class is installed and all subsequent years in $I^{\text{exdev}}_{b,u}$.

We introduce non-negative continuous variables $\dot{E}^{\text{avail}}_{b,u,i}$ that represent the available device size of device $u \in U$ to be used in building $b \in B$ and year $i \in I$. Constraints (3.26) and (3.27) calculate values of $\dot{E}^{\text{avail}}_{b,u,i}$ through summation over variables $\dot{E}^{\text{dev}}_{b,u,j,i}$ for time steps $j \in I^{\text{avail}}_{u,i}$ while considering potentially existing devices before the optimization time horizon.

Figure 3.4 exemplarily shows the mechanics of previously introduced constraints (3.21)–(3.27) for device $\tilde{u} \in U \setminus U^{\text{roof}}$ and building $\tilde{b} \in B$ with $l^{\text{dev}}_{\tilde{u}} = 4$ and $l^{\text{exdev}}_{\tilde{b},\tilde{u}} = 2$. Without loss of generality, we assume that $\tilde{u}$ may be the only device of its technology class. In Fig. 3.4, dashed arrows relate to values of $\dot{E}^{\text{dev}}_{b,u,i,j}$ that are forced to a value of zero by attached constraints, while solid lines relate to variables $\dot{E}^{\text{dev}}_{b,u,i,j}$ that may take values between zero and $o^{\text{dev}}_{b,u}$. If device $\tilde{u}$ is only installed in years $i = 1$ and $i = 3$, i.e., $X^{\text{dev}}_{\tilde{b},\tilde{u},1} = X^{\text{dev}}_{\tilde{b},\tilde{u},3} = 1$ and $X^{\text{dev}}_{\tilde{b},\tilde{u},0} = X^{\text{dev}}_{\tilde{b},\tilde{u},2} = X^{\text{dev}}_{\tilde{b},\tilde{u},4} = 0$, the existing device can be used only in $i = 0$ since the value of $\dot{E}^{\text{exdev}}_{\tilde{b},\tilde{u},1}$ is forced to value zero by means of constraints (3.25). Similarly, for years $i \in \{0, 2, 4\}$ in which device $\tilde{u}$ is not installed, variables $\dot{E}^{\text{dev}}_{b,u,i,j}$ are forced to value zero by means of constraints (3.22). Variables $\dot{E}^{\text{dev}}_{b,u,1,j}$ may take positive values for $j \in \{1, 2\}$ because of constraints (3.21) but are forced to zero for $j = 3$ by constraints (3.22) and for $j = 4$ by constraints (3.22) and (3.23). The calculation of variables $\dot{E}^{\text{avail}}_{b,u,i}$ follows with constraints (3.26) and (3.27) for $i \in I^{\text{exdev}}_{\tilde{b},\tilde{u}} = \{0, 1\}$ and $i \in I \setminus I^{\text{exdev}}_{\tilde{b},\tilde{u}} = \{2, \dots, n^{\text{years}} - 1\}$, respectively.

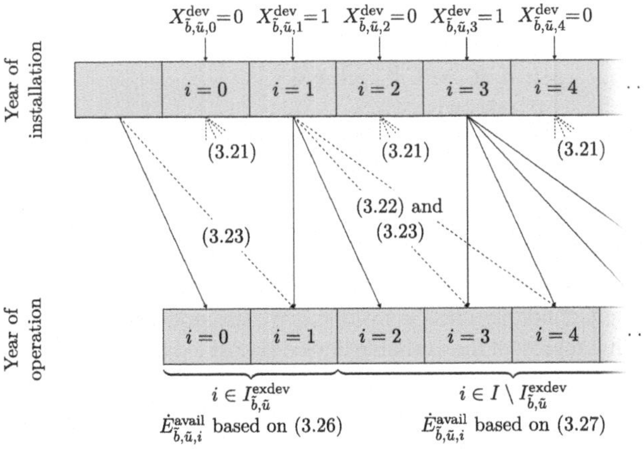

Fig. 3.4 Modeling mechanics of constraints (3.21)–(3.27) and relations between years of installation and operation for a device $\tilde{u} \in U \setminus U^{\text{roof}}$ and building $\tilde{b} \in B$. Dashed arrows relate to values of $\dot{E}^{\text{dev}}_{b,u,i,j}$ that are forced to a value of zero by attached constraints, while solid lines relate to variables $\dot{E}^{\text{dev}}_{b,u,i,j}$ that may take values between zero and $o^{\text{dev}}_{b,u}$

To ensure that peak demand $p \in P$ is covered in each year $i \in I$ for buildings $b \in B$, we use constraints (3.28) to ensure that the sum of device sizes $\dot{E}^{\text{avail}}_{b,u,i}$ for heat generating devices $u \in U^{\text{heat}}_p$ that may cover $p^{\text{peak}}_{b,p}$ is large enough. Lastly, for devices $u \in U^{\text{roof}}$ we need to ensure that the maximum roof area a^{roof}_b of building $b \in B$ is not exceeded. For devices $u \in U^{\text{roof}}$, we use variables $\dot{E}^{\text{dev}}_{b,u,i,i}$ to represent the installed area in building $b \in B$ and year $i \in I$. Lastly, constraints (3.29) ensure that for every building $b \in B$ and year $i \in I$, the available roof area $a^{\text{avail}}_{b,i}$ is not exceeded. The available roof area is given by

$$a_{b,i}^{\text{avail}} = a_b^{\text{roof}} - \sum_{\substack{u \in U^{\text{roof}}: \\ i \in I_{b,u}^{\text{exdev}}}} a_{b,u}^{\text{roofdev}} \qquad \forall b \in B, i \in I \tag{3.34}$$

and thus respects the roof area $a_{b,u}^{\text{roofdev}}$ already covered by existing devices $u \in U^{\text{roof}}$ in $b \in B$.

3.3.4 Building Operations

$$\dot{E}_{b,u,i,d,\tau}^{\text{op}} \leq \min_{f \in H_u^{\text{out}}} \{\eta_{b,u,f,d,\tau}\}^{-1} \dot{E}_{b,u,i}^{\text{avail}} \quad \forall b \in B, u \in U \setminus (U^{\text{sto}} \cup U^{\text{roof}}), i \in I, d \in D, \tau \in T \tag{3.35}$$

$$\dot{E}_{b,u,i,d,\tau}^{\text{op}} \leq g_{b,d,\tau} \dot{E}_{b,u,i}^{\text{avail}} \qquad \forall b \in B, u \in U^{\text{roof}}, i \in I, d \in D, \tau \in T \tag{3.36}$$

$$SOC_{b,u,i,d,\tau} \leq dod_u \dot{E}_{b,u,i}^{\text{avail}} \qquad \forall b \in B, u \in U^{\text{sto}}, i \in I, d \in D, \tau \in T \tag{3.37}$$

$$CH_{b,u,i,d,\tau} \leq \frac{dod_u}{\eta_u^{\text{ch}} \Delta_u^{\text{sto}}} \dot{E}_{b,u,i}^{\text{avail}} \qquad \forall b \in B, u \in U^{\text{sto}}, i \in I, d \in D, \tau \in T \tag{3.38}$$

$$DCH_{b,u,i,d,\tau} \leq \frac{dod_u}{\Delta_u^{\text{sto}}} \dot{E}_{b,u,i}^{\text{avail}} \qquad \forall b \in B, u \in U^{\text{sto}}, i \in I, d \in D, \tau \in T \tag{3.39}$$

$$\begin{aligned} \dot{E}_{b,f,i,d,\tau}^{\text{dem}} = & \sum_{\substack{u \in U \setminus U^{\text{sto}}: \\ f \in H_u^{\text{out}}}} \eta_{b,u,f,d,\tau} \dot{E}_{b,u,i,d,\tau}^{\text{op}} - \sum_{\substack{u \in U \setminus U^{\text{sto}}: \\ f \in H_u^{\text{in}}}} \dot{E}_{b,u,i,d,\tau}^{\text{op}} \\ & + \sum_{\substack{u \in U^{\text{sto}}: \\ f \in H_u^{\text{out}}}} \left(\eta_u^{\text{dch}} DCH_{b,u,i,d,\tau} - CH_{b,u,i,d,\tau} \right) \\ & + \sum_{\substack{g \in G_b^{\text{imp}}: \\ g = f}} \dot{E}_{b,g,i,d,\tau}^{\text{imp}} - \sum_{\substack{g \in G^{\text{exp}}: \\ g = f}} \dot{E}_{b,g,i,d,\tau}^{\text{exp}} \\ & \forall b \in B, f \in F, i \in I, d \in D, \tau \in T \end{aligned} \tag{3.40}$$

$$\dot{E}_{b,\text{th},i,d,\tau}^{\text{dem}} = \sum_{e \in E} p_{b,e,d,\tau}^{\text{heat}} Y_{b,e,i}^{\text{eff}} + p_{b,d,\tau}^{\text{dhw}} \qquad \forall b \in B, i \in I, d \in D, \tau \in T \tag{3.41}$$

$$\dot{E}_{b,\text{el},i,d,\tau}^{\text{dem}} = p_{b,d,\tau}^{\text{el}} \qquad \forall b \in B, i \in I, d \in D, \tau \in T \tag{3.42}$$

$$\dot{E}_{b,g,i,d,\tau}^{\text{imp}} = \sum_{\substack{u \in U \setminus U^{\text{sto}}: \\ g \in H_u^{\text{in}}}} \dot{E}_{b,u,i,d,\tau}^{\text{op}} \qquad \forall b \in B, g \in G_b^{\text{imp}} : g \notin F, i \in I, d \in D, \tau \in T \tag{3.43}$$

$$\begin{aligned} SOC_{b,u,i,d,\tau} = {} & (1 - el_u)\, SOC_{b,u,i,d,\tau-1} \\ & - \Delta^{\text{ts}} DCH_{b,u,i,d,\tau} \\ & + \Delta^{\text{ts}} \eta_u^{\text{ch}} CH_{b,u,i,d,\tau} \qquad \forall b \in B, u \in U^{\text{sto}}, i \in I, d \in D, \tau \in T \setminus \{0\} \end{aligned} \tag{3.44}$$

$$\begin{aligned} SOC_{b,u,i,d,0} = {} & (1 - el_u)\, SOC_{b,u,i,d,|T|-1} \\ & - \Delta^{\text{ts}} DCH_{b,u,i,d,0} \\ & + \Delta^{\text{ts}} \eta_u^{\text{ch}} CH_{b,u,i,d,0} \qquad \forall b \in B, u \in U^{\text{sto}}, i \in I, d \in D \end{aligned} \tag{3.45}$$

$$\begin{aligned} 0 \geq {} & \tilde{t}^{\text{dhw}} p_{b,d,\tau}^{\text{dhw}} \\ & - \sum_{u \in U_{\text{dhw}}^{\text{heat}}} \eta_{b,u,\text{th},d,\tau} \dot{E}_{b,u,i,d,\tau}^{\text{op}} \qquad \forall b \in B : \text{th} \notin G_b^{\text{imp}}, i \in I, \\ & d \in D, \tau \in T \end{aligned} \tag{3.46}$$

$$\begin{aligned} \dot{E}_{b,\text{th},i,d,\tau}^{\text{imp}} \geq {} & \tilde{t}^{\text{dhw}} p_{b,d,\tau}^{\text{dhw}} \\ & - \sum_{u \in U_{\text{dhw}}^{\text{heat}}} \eta_{b,u,\text{th},d,\tau} \dot{E}_{b,u,i,d,\tau}^{\text{op}} \qquad \forall b \in B : \text{th} \in G_b^{\text{imp}}, i \in I, \\ & d \in D, \tau \in T \end{aligned} \tag{3.47}$$

$$SOC_{b,u,i,d,\tau} \in \mathbb{R}_{\geq 0} \qquad \forall b \in B, u \in U, i \in I, d \in D, \tau \in T \tag{3.48}$$

$$CH_{b,u,i,d,\tau} \in \mathbb{R}_{\geq 0} \qquad \forall b \in B, u \in U, i \in I, d \in D, \tau \in T \tag{3.49}$$

$$DCH_{b,u,i,d,\tau} \in \mathbb{R}_{\geq 0} \qquad \forall b \in B, u \in U, i \in I, d \in D, \tau \in T \tag{3.50}$$

$$\dot{E}_{b,u,i,d,\tau}^{\text{op}} \in \mathbb{R}_{\geq 0} \qquad \forall b \in B, u \in U \setminus U^{\text{sto}}, i \in I, d \in D, \tau \in T \tag{3.51}$$

$$\dot{E}_{b,f,i,d,\tau}^{\text{dem}} \in \mathbb{R}_{\geq 0} \qquad \forall b \in B, f \in F, i \in I, d \in D, \tau \in T \tag{3.52}$$

$$\dot{E}_{b,g,i,d,\tau}^{\text{exp}} \in \mathbb{R}_{\geq 0} \qquad \forall b \in B, g \in G^{\text{exp}}, i \in I, d \in D, \tau \in T \tag{3.53}$$

$$\dot{E}_{b,g,i,d,\tau}^{\text{imp}} \in \mathbb{R}_{\geq 0} \qquad \forall b \in B, g \in G_b^{\text{imp}}, i \in I, d \in D, \tau \in T \tag{3.54}$$

To model operations of building energy systems, we introduce non-negative continuous variables $\dot{E}_{b,u,i,d,\tau}^{\text{op}}$ that represent the energy inflow rate of device $u \in U \setminus U^{\text{sto}}$ in building $b \in B$ and year $i \in I$ on typical day $d \in D$ and time step $\tau \in T$. For devices $u \in U \setminus (U^{\text{sto}} \cup U^{\text{roof}})$, constraints (3.35) ensure that the energy outflow rate is never larger than $\dot{E}_{b,u,i}^{\text{avail}}$ which represents the currently installed and usable device size of $u \in U$ in $b \in B$ and $i \in I$. In constraints (3.35), we use $\min_{f \in H_u^{\text{out}}} \{\eta_{b,u,f,d,\tau}\}$ to ensure for devices with $|H_u^{\text{out}}| \geq 2$, e.g., CHP, that the output energy carrier with the lowest device efficiency determines the maximum energy inflow rate of the device to ensure the maximum energy outflow rate is restricted by $\dot{E}_{b,u,i}^{\text{avail}}$ for energy carrier $\arg\min_{f \in H_u^{\text{out}}} \{\eta_{b,u,f,d,\tau}\}$. For devices $u \in U^{\text{roof}}$, constraints (3.36) model that the maximum energy inflow rate $\dot{E}^{\text{op}}_{b,u,i,d,\tau}$ is restricted by the covered roof area $\dot{E}^{\text{avail}}_{b,u,i}$ multiplied by the normal solar irradiation $g_{b,d,\tau}$. Therefore, the maximum energy inflow rate $\dot{E}^{\text{op}}_{b,u,i,d,\tau}$ is given by the total normal solar irradiation on device u in building $b \in B$ and is reduced to the energy outflow rate of $f \in H_u^{\text{out}}$ by means of efficiency $\eta_{b,u,f,d,\tau}$. For devices $u \in U^{\text{sto}}$, we further introduce non-negative continuous variables $SOC_{b,u,i,d,\tau}$, $CH_{b,u,i,d,\tau}$ and $DCH_{b,u,i,d,\tau}$ that represent the state-of-charge (SOC), the charging rate and discharging rate for device $u \in U^{\text{sto}}$ in building $b \in B$ and year $i \in I$ on typical day $d \in D$ and time step $\tau \in T$, respectively. In constraints (3.37), we limit the maximum SOC to the available storage size $\dot{E}^{\text{avail}}_{b,u,i}$ times the DOD dod_u of the storage. Furthermore, constraints (3.38) and (3.39) limit the maximum rate of charge and discharge for storage $u \in U^{\text{sto}}$ in building $b \in B$ and year $i \in I$ on typical day $d \in D$ and time step $\tau \in T$.

For final energy carriers $f \in F$, we need to ensure that energy demand is met at all times. We introduce non-negative continuous variables $\dot{E}^{\text{dem}}_{b,f,i,d,\tau}$ that represent the demanded energy flow rate in building $b \in B$ for final energy carrier $f \in F$ in year $i \in I$ on typical day $d \in D$ and at time step $\tau \in T$. Furthermore, we define non-negative continuous variables $\dot{E}^{\text{exp}}_{b,g,i,d,\tau}$ for $g \in G^{\text{exp}}$ and $\dot{E}^{\text{imp}}_{b,g,i,d,\tau}$ for $g \in G_b^{\text{imp}}$ to denote the energy flow rates exported to or imported from the grid, respectively. Then, in constraints (3.40), we use quasi-stationary energy balances for the energy distribution systems that connect respective energy demands, installed devices and the grid (cf. Fig. 3.3). The quasi-stationary energy balance states that at every time step, entering and leaving energy flow rates of devices, grid, and demand need to be equally high. The assumption of quasi-stationarity is not expected to introduce significant modeling errors, as for typical time step lengths, the change in internal energy of the thermodynamic system is relatively small compared to the energy flow across the system boundary. With constraints (3.41), we state that the energy flow rate for $f = \text{th}$ equates the heat demand for DHW plus the heat demand for space heating that depends on the building efficiency $e \in E$. Similarly, we use constraints (3.42) to specify the building demand for $f = \text{el}$. For energy carriers $g \in G_b^{\text{imp}} : g \notin F$, i.e., energy carriers from the grid that are not final energy carriers, e.g., gas, the imported energy flow rate $\dot{E}^{\text{imp}}_{b,g,i,d,\tau}$ is calculated with the quasi-stationary energy balance in constraints (3.43) for devices $u \in U \setminus U^{\text{sto}}$ with $g \in H_u^{\text{in}}$. Note that the imported energy flow rates $\dot{E}^{\text{imp}}_{b,g,i,d,\tau}$ for $g \in G_b^{\text{imp}} : g \in F$ are determined by constraints (3.40).

We model storages with integrated internal energy balances in constraints (3.44) and (3.45) where we consider energy losses that are proportional to the SOC of the previous time step $SOC_{b,u,u,i,d,\tau-1}$ by a factor of el_u. The integrated internal energy balance of a storage $u \in U^{\text{sto}}$ is depicted in Fig. 3.5. Moreover, constraints (3.45) ensure that the SOC before the first time step of a typical day equals the SOC after the last time step of the same day. Therefore, energy cannot be carried over to other typical days of the same year. Lastly, we introduce constraints (3.46) to ensure that at any given time step, the aggregate heat flow rate of devices $u \in U^{\text{heat}}_{\text{dhw}}$ in buildings $b \in B$: th $\notin G^{\text{imp}}_b$ is high enough to heat DHW to a temperature of t^{dhw}. For buildings $b \in B$: th $\in G^{\text{imp}}_b$, constraints (3.47) ensure that DHW may be heated up to a temperature of t^{dhw} while also considering district heating.

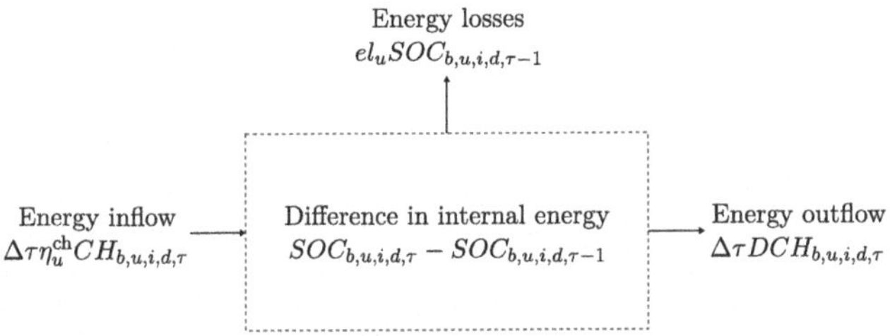

Fig. 3.5 Integrated internal energy balance for storage devices $u \in U^{\text{sto}}$. The change in internal energy is due to the difference in energy inflows and energy outflows for a given time step Δ^{ts} as well as the energy loss that is proportional to the state of charge $SOC_{b,u,i,d,\tau-1}$ by a factor of el_u. The energy inflow $\Delta^{\text{ts}}\eta_u^{\text{ch}} CH_{b,u,i,d,\tau}$ contains storage charging efficiency η_u^{ch}, as $CH_{b,u,i,d,\tau}$ is an energy flow rate leaving the system boundaries of energy balance (3.40)

3.3.5 Objective Function

As previously mentioned in the problem description, the problem asks for an evaluation of the trade-off between economic and ecological aspects of transformation pathways in the residential building sector. To model this trade-off, we use a weighted objective function that facilitates the combined optimization of both optimization goals. We choose NPV and TGW as economical and ecological single objective functions, respectively. Both the NPV and TGW are impacted by the transformation pathways of the residential building stock and the operations of its building energy systems. To facilitate such an optimization, we minimize a convex combination of the negative NPV and TGW multiplied by a constant of proportionality, which leaves us with the objective function

$$\min \quad -\phi\, NPV + (1-\phi)\,\omega\, TGW \tag{3.55}$$

where both NPV and TGW are continuous variables. The parameter $\phi \in [0, 1]$ is used for the convex combination of $-NPV$ and ωTGW. The constant of proportionality $\omega > 0$ in (3.55) is used to scale the TGW in (3.55) appropriately to bring its contribution to the objective function into a similar order of magnitude of the NPV. In the following, we present constraints to determine the NPV and TGW. Note that we disaggregate the calculation of various components of NPV and TGW to state constraints directly usable within our model decompositions in Chap. 4.

$$NPV = -\left(C^{\text{inv}} + C^{\text{om}} + C^{\text{en}}\right) \tag{3.56}$$

$$C^{\text{inv}} = C^{\text{inv,dev,fix}} + \sum_{b\in B}\sum_{i\in I} C^{\text{inv,dev,var}}_{b,i} + C^{\text{inv,mea}} \tag{3.57}$$

$$C^{\text{inv,dev,fix}} = \sum_{b\in B}\sum_{i\in I}\sum_{u\in U} \frac{w^{\text{buil}}_b\left(1-h^{\text{dev}}_{u,i}\right)}{(1+\rho)^{i+1}} c^{\text{dev,fix}}_{u,i} X^{\text{dev}}_{b,u,i} \tag{3.58}$$

$$C^{\text{inv,dev,var}}_{b,i} = \sum_{u\in U} \frac{w^{\text{buil}}_b\left(1-h^{\text{dev}}_{u,i}\right)}{(1+\rho)^{i+1}} c^{\text{dev,var}}_{u,i} \dot{E}^{\text{dev}}_{b,u,i,i} \qquad \forall b \in B, i \in I \tag{3.59}$$

$$C^{\text{inv,mea}} = \sum_{b\in B}\sum_{m\in M}\sum_{i\in I} \frac{w^{\text{buil}}_b\left(1-h^{\text{mea}}_{m,i}\right)}{(1+\rho)^{i+1}} c^{\text{mea}}_{b,m,i} X^{\text{mea}}_{b,m,i} \tag{3.60}$$

$$C^{\text{om}} = C^{\text{om,fix}} + \sum_{b\in B}\sum_{i\in I} C^{\text{om,dev,var}}_{b,i} \tag{3.61}$$

$$C^{\text{om,fix}} = \sum_{b\in B}\sum_{u\in U} w^{\text{buil}}_b\Bigg(\sum_{i\in I^{\text{exdev}}_{b,u}} C^{\text{om,exdev}}_{b,u,i} + \sum_{i\in I}\sum_{j\in I^{\text{dev}}_{u,i}} C^{\text{om,dev,fix}}_{b,u,i,j}\Bigg) \tag{3.62}$$

$$C^{\text{om,exdev}}_{b,u,i} \geq \frac{c^{\text{om}}_u c^{\text{exdev}}_{b,u}}{(1+\rho)^{i+1}}\left(1 - \sum_{v\in U}\sum_{j=0}^{i} z^{\text{dev}}_{u,v} X^{\text{dev}}_{b,v,j}\right) \qquad \forall b \in B, u \in U, i \in I^{\text{exdev}}_{b,u} \tag{3.63}$$

$$C^{\text{om,dev,fix}}_{b,u,i,j} \geq \frac{c^{\text{om}}_u c^{\text{dev,fix}}_{u,i}}{(1+\rho)^{j+1}}\left(X^{\text{dev}}_{b,u,i} - \sum_{v\in U}\sum_{\tilde{i}=i+1}^{j} z^{\text{dev}}_{u,v} X^{\text{dev}}_{b,v,\tilde{i}}\right) \qquad \forall b \in B, u \in U, i \in I, j \in I^{\text{dev}}_{u,i} \tag{3.64}$$

$$C^{\text{om,dev,var}}_{b,i} \geq \sum_{u\in U} \frac{w^{\text{buil}}_b c^{\text{om}}_u}{(1+\rho)^{i+1}} \sum_{j\in I^{\text{avail}}_{u,i}} c^{\text{dev,var}}_{u,j} \dot{E}^{\text{dev}}_{b,u,j,i} \qquad \forall b \in B, i \in I \tag{3.65}$$

$$C^{\text{en}} = \sum_{b\in B}\sum_{i\in I} C^{\text{en,sp}}_{b,i} \tag{3.66}$$

$$C^{\text{en,sp}}_{b,i} = \sum_{d\in D}\sum_{\tau\in T} \frac{w^{\text{buil}}_b w^{\text{day}}_{b,d} \Delta^{\text{ts}}}{(1+\rho)^{i+1}} \Big(\sum_{g\in G^{\text{imp}}_b} c^{\text{en}}_{g,i} \dot{E}^{\text{imp}}_{b,g,i,d,\tau} - \sum_{g\in G^{\text{exp}}} s^{\text{en}}_{g,i} \dot{E}^{\text{exp}}_{b,g,i,d,\tau} \Big) \quad \forall b \in B, i \in I \tag{3.67}$$

$$TGW = \sum_{b\in B}\sum_{i\in I} TGW^{\text{op}}_{b,i} + TGW^{\text{mea}} \tag{3.68}$$

$$TGW^{\text{op}}_{b,i} = \sum_{d\in D}\sum_{\tau\in T} w^{\text{buil}}_b w^{\text{day}}_{b,d} \Delta^{\text{ts}} \Big(\sum_{g\in G^{\text{imp}}_b} f^{\text{en}}_{g,i} \dot{E}^{\text{imp}}_{b,g,i,d,\tau} - \sum_{g\in G^{\text{exp}}} f^{\text{en}}_{g,i} \dot{E}^{\text{exp}}_{b,g,i,d,\tau} \Big) \quad \forall b \in B, i \in I \tag{3.69}$$

$$TGW^{\text{mea}} = \sum_{b\in B}\sum_{m\in M}\sum_{i\in I} w^{\text{buil}}_b f^{\text{mea}}_{b,m} X^{\text{mea}}_{b,m,i} \tag{3.70}$$

$$NPV \in \mathbb{R} \tag{3.71}$$

$$TGW \in \mathbb{R} \tag{3.72}$$

$$C^{\text{inv}} \in \mathbb{R}_{\geq 0} \tag{3.73}$$

$$C^{\text{inv,dev,fix}} \in \mathbb{R}_{\geq 0} \tag{3.74}$$

$$C^{\text{inv,dev,var}}_{b,i} \in \mathbb{R}_{\geq 0} \quad \forall b \in B, i \in I \tag{3.75}$$

$$C^{\text{inv,mea}} \in \mathbb{R}_{\geq 0} \tag{3.76}$$

$$C^{\text{om}} \in \mathbb{R}_{\geq 0} \tag{3.77}$$

$$C^{\text{om,fix}} \in \mathbb{R}_{\geq 0} \tag{3.78}$$

$$C^{\text{om,dev,fix}}_{b,u,i,j} \in \mathbb{R}_{\geq 0} \quad \forall b \in B, u \in U, i \in I, j \in I^{\text{dev}}_{u,i} \tag{3.79}$$

$$C^{\text{om,exdev}}_{b,u,i} \in \mathbb{R}_{\geq 0} \quad \forall b \in B, u \in U, i \in I^{\text{exdev}}_{b,u} \tag{3.80}$$

$$C^{\text{om,dev,var}}_{b,u,i} \in \mathbb{R}_{\geq 0} \quad \forall b \in B, u \in U, i \in I \tag{3.81}$$

$$C^{\text{en}} \in \mathbb{R} \tag{3.82}$$

$$C^{\text{en,sp}}_{b,i} \in \mathbb{R} \quad \forall b \in B, i \in I \tag{3.83}$$

$$TGW^{\text{op}}_{b,i} \in \mathbb{R} \quad \forall b \in B, i \in I \tag{3.84}$$

$$TGW^{\text{mea}} \in \mathbb{R}_{\geq 0} \tag{3.85}$$

For the calculation of the value of NPV with constraint (3.56), we introduce non-negative continuous variables C^{inv} and C^{om} as well as the continuous variable C^{en}

to denote discounted costs of investment, discounted costs of O&M and discounted costs of energy, respectively. The value of NPV is then defined as the negative sum of C^{inv}, C^{om} and C^{en}, cf. constraint (3.56).

Constraint (3.57) states that discounted costs of investment C^{inv} comprise discounted fixed and variable investment costs into devices and discounted investment costs of efficiency measures, for which we respectively introduce non-negative continuous variables $C^{\text{inv,dev,fix}}$, $C^{\text{inv,dev,var}}_{b,i}$ and $C^{\text{inv,mea}}$. Discounted fixed costs for investments $C^{\text{inv,dev,fix}}$ are given by constraint (3.58). Analogously to Richarz et al. (2022), we use an auxiliary parameter $h^{\text{dev}}_{u,i}$ in (3.58) to account for the residual value of devices at the end of the optimization time horizon. Thereby, the parameter is defined by the relationship

$$h^{\text{dev}}_{u,i} = \begin{cases} \dfrac{l^{\text{dev}}_u - (n^{\text{years}} - i)}{l^{\text{dev}}_u \, (1+\rho)^{n^{\text{years}}-i}} & \text{if } l^{\text{dev}}_u > n^{\text{years}} - i \\ 0 & \text{otherwise} \end{cases} \qquad \forall u \in U, i \in I. \tag{3.86}$$

Variables $C^{\text{inv,dev,var}}_{b,i}$ denote discounted costs for the investment into devices proportional to the installed device size $\dot{E}^{\text{avail}}_{b,u,i,i}$. The values of variables $C^{\text{inv,dev,var}}_{b,i}$ are determined by constraints (3.59). Note that we disaggregate the calculation of discounted investment costs of devices into constraints (3.64) and (3.59) to facilitate the development of model decompositions for Benders decomposition in Chap. 4. Discounted costs of investments into efficiency measure C^{mea} are given by constraint (3.60), where we transfer the concept of calculating the residual values for efficiency measures by introducing auxiliary parameters $h^{\text{mea}}_{m,i}$ defined by

$$h^{\text{mea}}_{m,i} = \begin{cases} \dfrac{l^{\text{mea}}_m - (n^{\text{years}} - i)}{l^{\text{mea}}_m \, (1+\rho)^{n^{\text{years}}-i}} & \text{if } l^{\text{mea}}_m > n^{\text{years}} - i \\ 0 & \text{otherwise} \end{cases} \qquad \forall m \in M, i \in I. \tag{3.87}$$

For the calculation of discounted costs of O&M, C^{om}, we introduce non-negative continuous variables $C^{\text{om,fix}}$ and $C^{\text{om,dev,var}}_{b,i}$. Values for $C^{\text{om,fix}}$ are calculated using constraint (3.62) and comprise discounted costs for O&M of existing devices and discounted fixed costs for O&M proportional to fixed investment costs for installed devices, both represented by newly introduced non-negative continuous variables $C^{\text{om,exdev}}_{b,u,i}$ and $C^{\text{om,dev,fix}}_{b,u,i,j}$. The calculation of variables $C^{\text{om,exdev}}_{b,u,i}$ is achieved by inequalities (3.63) that account for a proportional share of the original total investment costs $c^{\text{exdev}}_{b,u}$ of existing device u with factor c^{om}. For a given year $i \in I$, constraints (3.63) set the lower bound on $C^{\text{om,exdev}}_{b,u,i}$ to zero if a device of the same technology class is installed in any year $j = \{0, \ldots, i\}$. Similarly to the modeling mechanics used in constraints (3.63), we account for discounted costs for O&M due to the fixed share of investment costs with constraints (3.64). Discounted costs $C^{\text{om,dev,fix}}_{b,u,i,j}$ calculated by (3.64) account for costs for O&M in year $j \in I^{\text{dev}}_{u,i}$ of a device $u \in U$ installed in year $i \in I$ and building $b \in B$. Therefore, in particular, constraints (3.64) account for changes in fixed device costs $c^{\text{dev,fix}}_{u,i}$. If in year $i \in I$ device $u \in U$ is installed in building $b \in B$, constraints (3.64) set a lower bound of $c^{\text{om}}_u c^{\text{dev,fix}}_{u,i} / (1+\rho)^{j+1}$ on variables $C^{\text{om,dev,fix}}_{b,u,i,j}$ for $j \in I^{\text{dev}}_{u,i}$ for as long as no other device of the same technology class is installed. Lastly, we calculate discounted costs for O&M $C^{\text{om,dev,var}}_{b,i}$ due to the variable share of investment costs with constraints (3.65). Here, we follow the aforementioned assumption that O&M costs due to the variable share of investment costs are set proportional to the used device size $\dot{E}^{\text{dev}}_{b,u,j,i}$ for device $u \in U$ installed in year $j \in I$. Note that we disaggregate the calculation of discounted costs for O&M of new devices with constraints (3.61)–(3.65) to facilitate the Benders decomposition in Chap. 4.

Discounted costs of energy for building energy system operations C^{en} are calculated using constraints (3.66) for which we introduce continuous variables $C^{\text{en,sp}}_{b,i}$ that represent discounted costs for building energy system operations for building $b \in B$ and year $i \in I$. The values of variables $C^{\text{en,sp}}_{b,i}$ are determined by constraints (3.67) and depend on the aggregated amounts of imported and exported energy, respectively. Note that we disaggregate the calculation of C^{en} with constraints (3.66) and (3.67) to facilitate the Benders decomposition in Chap. 4.

Constraint (3.68) ensures the calculation of TGW. We introduce continuous variables $TGW^{\text{op}}_{b,i}$ and the non-negative continuous variable TGW^{mea} that denote the share of total climate change due to the operation of the building energy system of building $b \in B$ in year $i \in I$ as well as the share of TGW due to aggregated emissions embodied in efficiency measures of the entire building stock. Values for $TGW^{\text{op}}_{b,i}$ are calculated with constraints (3.69) and depend on aggregated amounts of imported and exported energy and the emission factor $f^{\text{en}}_{g,i}$ of energy carrier $g \in G^{\text{imp}}_b$ or $g \in G^{\text{exp}}$ in year $i \in I$. Lastly, we calculate TGW^{mea} related to the realization of efficiency measures with constraint (3.70). Note that we disaggregate the calculation of TGW with constraints (3.68)–(3.70) to facilitate the Benders decomposition in Chap. 4.

Model Decompositions

4

In this chapter, we present different model decompositions of the MILP model introduced in Chap. 3 that are then used within a Benders decomposition algorithm developed in this work. To deduce different model decompositions, we first analyze the model structure of the introduced MILP model in Sect. 4.1. Based on the identified model structure, we propose a modular decomposition strategy that allows for decomposing the model by using the previously introduced model structure of the problem. Subsequently, we introduce different model modifications in Sect. 4.2 that aim to reduce the time needed for the generation of Benders cuts after solving Benders SPs. Lastly, in Sect. 4.3, we state the model decompositions alongside a set of valid inequalities to be used in Benders RMPs as well as a cut strengthening technique that can be employed for Benders optimality cuts in some of the proposed decompositions.

4.1 Model Structure and Modular Decomposition Strategy

In the following, we will analyze the structure of the model proposed in Chap. 3 to identify possible model decompositions. Figure 4.1 depicts the coefficient matrix of the proposed model where the first rows are constraints pertaining to the resource allocation at the building stock level and the following rows are constraints grouped by single buildings $b \in B$, following the problem structure presented in Sect. 3.1. The first block in Fig. 4.1 containing constraints pertaining to resource allocation at building stock level link binary variables $X^{\text{dev}}_{b,u,i}$ and $X^{\text{mea}}_{b,u,i}$ across buildings $b \in B$ for respective years $i \in I$. Furthermore, this block contains variables $Y^{\text{encon}}_{b,i}$ that are used to restrict the maximum number of energy consultancies. Next to the first block, $X^{\text{dev}}_{b,u,i}$ also appear in constraints determining building energy

R. Delorme, *Optimization of Transformation Pathways for the Residential Building Sector*, BestMasters, https://doi.org/10.1007/978-3-658-47466-9_4

system pathways of buildings. Therefore, variables $X^{\text{dev}}_{b,u,i}$ represent linking variables between constraints describing resource allocation at the building stock level and constraints describing energy system pathways. Similarly, variables $X^{\text{mea}}_{b,u,i}$ appear in constraints to determine building efficiency pathways for each building and thus link those blocks to the building stock allocation., cf. Fig. 4.1. In constraints pertaining to building operations, variables $Y^{\text{eff}}_{b,e,i}$ link the state of building efficiency

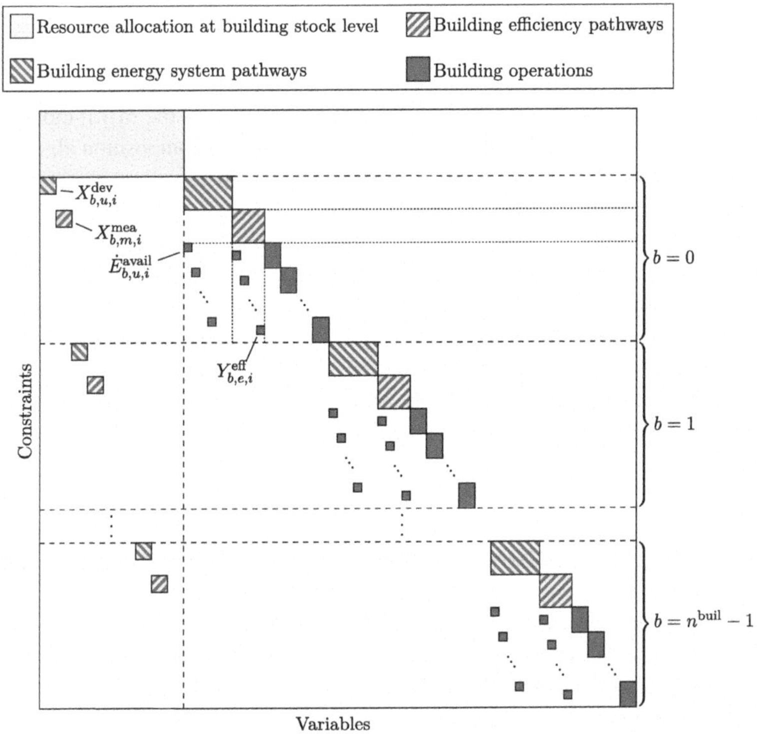

Fig. 4.1 Structure of the coefficient matrix of the MILP model. Constraints in the master block pertain to the resource allocation on the building stock level. The following rows are constraints grouped by single buildings $b \in B$, following the problem structure presented in Sect. 3.1. Variables $X^{\text{dev}}_{b,u,i}$ and $X^{\text{mea}}_{b,m,i}$ link constraints for resource allocation at the building stock level with building energy system and building efficiency pathways, respectively. Similarly, variables $\dot{E}^{\text{avail}}_{b,u,i}$ and $Y^{\text{eff}}_{b,e,i}$ respectively link building energy system and building efficiency pathways with building operations. Depicted geometric relations do not correspond to the true relations between the numbers of constraints and variables of the different model parts

of each year to the building operations of that year. Analogously, variables $\dot{E}^{\text{avail}}_{b,u,i}$ link the respective energy system design, as a state within building energy system pathways, to building operations of the respective years. Note that constraints of building operations can therefore be decoupled by year. Furthermore, those constraints could even be coupled by typical days, which, however, is not depicted in Fig. 4.1 for enhanced overview.

As described, the three model levels and their four parts introduced in Chap. 3 are linked by means of variables $X^{\text{dev}}_{b,u,i}$, $X^{\text{mea}}_{b,m,i}$, $Y^{\text{eff}}_{b,e,i}$ and $\dot{E}^{\text{avail}}_{b,u,i}$. In the following, depending on the respective decomposition, we will call those variables *linking variables* if they are part of both the Benders RMP and Benders SPs. Based on the structure of the coefficient matrix depicted in Fig. 4.1, we decide to group constraints to be added to Benders SPs based on the presented problem structure. Since in resource allocation at the building stock level we use binary variables to decide on the building $b \in B$ and year $i \in I$ for the installation of devices $u \in U$ or realization of efficiency measures $m \in M$, we choose to move corresponding constraints and variables to the Benders RMP in all our decompositions. Conversely, building operations depend on both building efficiency pathways and building energy system pathways. Therefore, we use constraints pertaining to building operations to Benders SPs in all our decompositions. The decompositions, therefore, differ in whether constraints pertaining to building efficiency pathways and building energy system pathways are moved to the Benders RMP or Benders SPs, respectively. Therefore, in the following, we will develop four different decompositions and name them based on the parts of the model that are included in Benders SPs:

1. BO – Building Operations
2. BOEP – Building Operations and Efficiency Pathways
3. BOESP – Building Operations and Energy System Pathways
4. BOESEP – Building Operations, Energy System and Efficiency Pathways

Note that, here, we limit ourselves to decompositions with only one single stage of Benders SPs. In a multi-stage or nested Benders decomposition setting, constraints of building operations could be moved to a second stage of Benders SPs to separate Benders cuts for Benders SPs of the first stage that comprise constraints pertaining to building efficiency pathways or building energy system pathways.

As depicted in Fig. 4.1, decomposition BO allows for the decomposition of Benders SPs for each building $b \in B$, year $i \in I$, and typical day $d \in D$, as for a given Benders RMP solution, the building efficiency and building energy system design is fixed and building operations are not linked in terms of typical days $d \in D$. Conversely, we observe that the Benders RMP is the largest in terms of the number of constraints compared to the other three decompositions. For all other decompositions BOEP, BOESP, and BOESEP, Benders SPs are only decomposable by buildings $b \in B$, as constraints of building efficiency pathways and building energy system pathways link years $i \in I$. However, the size of the Benders RMP is smaller in terms of the number of constraints compared to decomposition BO and is smallest for BOESEP.

When separating Benders cuts computationally, only the dual variables of a subset of constraints are relevant for the construction of a Benders cut. In the following, we present procedures to modify subsets of constraints of the developed model with the aim to accelerate the time needed for the Benders cut construction.

4.2 Model Modifications for Benders Cut Generation

As introduced in Sect. 2.3, Benders cuts comprise Benders feasibility cuts (2.26) and Benders optimality cuts (2.27). Without loss of generality, we now focus on the generation of Benders feasibility cuts with a single Benders SP to introduce procedures for model modifications with the aim to improve the computation time needed for Benders cut generation. However, the presented concepts are equally applicable to Benders optimality cuts and multiple Benders SPs. Furthermore, we refer to the primal problem (SP) as we use it for our implementation instead of the dual problem (DSP). If for a given solution of (RMP) $\bar{\mathbf{y}} \in \mathbb{Z}^t_{\geq 0}$ the primal problem (SP) is infeasible, we can retrieve an extreme ray $\boldsymbol{\pi}_r$ that proves the infeasibility of (SP), cf. Theorem 2. Furthermore, due to the structure of the Benders feasibility cut (2.26), we observe that we only need to consider dual values $\pi_{r,j}$ with $\pi_{r,j}$ being the j-th entry of the dual vector $\boldsymbol{\pi}_r$ where $b_j \neq 0$ or $b_{j,i} \neq 0$, with $b_{j,i}$ being the i-th entry of the j-th row of $\mathbf{B} \in \mathbb{R}^{s \times t}$ in (SP). Here, let $J = \{j \in \{1, \ldots, s\} : b_j \neq 0\} \cup \{j \in \{1, \ldots, s\} : \exists i \in \{1, \ldots, t\} : b_{j,i} \neq 0\}$. We observe that for the generation of Benders feasibility cuts, it is sufficient to check if $\pi_{r,j} \neq 0$ for rows $j \in J$. The left-hand-side (LHS) of Benders feasibility cuts is then generated by iteratively adding up terms $\pi_{r,j} \left(b_j - \sum_{i=1}^{t} b_{j,i} y_i\right)$ for which $\pi_{r,j} \neq 0$. Thus, it seems reasonable from a computational point of view, to seek a model formulation for Benders SPs with low $|J|$. Consequently, we apply the following two procedures to the constraints of the MILP model to derive optional model modifications to be used in Benders SPs:

(P1) For constraints with $b_j \neq 0$, i.e., a right-hand-side (RHS) unequal to zero, we replace concerned constraints with new constraints where we use a new variable instead of the RHS. Then, we introduce a new set of constraints to fix the value of that new variable to the RHS of the original constraint. Because of their purpose, we henceforth call those constraints *fixing constraints*. Instead of dual variables of original constraints with $b_j \neq 0$, the dual variables to be used for the construction of Benders cuts then correspond to the new fixing constraints. Therefore, if there are significantly more original constraints than fixing constraints, this procedure might lead to faster checking of dual variables and cut generation .

(P2) In constraints of the Benders SPs where linking variables have a matrix coefficient unequal to zero, i.e., constraints where $\exists i \in \{1, \ldots, t\} : b_{j,i} \neq 0$ for row j, we replace linking variables with new continuous variables (Codato and Fischetti, 2006, Stursberg, 2019, Maher, 2021). Then, we fix the new continuous variable to the original linking variables by means of another set of fixing constraints. Also, here, dual variables to be checked for Benders cut generation now correspond to the new fixing constraints.

Both procedures (P1) and (P2) increase the size of Benders SPs due to newly added constraints and variables at the potential expense of larger solving times. This potential increase in solving time might therefore overcompensate the reduction in time for cut generation. Therefore, we experimentally test the effectiveness of model modifications using procedures (P1) and (P2) in Chap. 6.

In the following, we show the precise modifications of constraints to be included in Benders SPs using procedures (P1) and (P2). We apply those procedures to the formed sets of constraints belonging to building efficiency pathways, building energy system pathways, and building operations. Table 4.1 shows the linking variables pertaining to each of the considered decompositions. Note that we only apply procedure (P2) to constraints in model decompositions where constraints contain linking variables of the Benders SPs. We do not modify constraints used for the calculation of costs or emissions, cf. Sect. 3.3.5, as we choose to respect objective coefficients of linking variables in the Benders RMP to avoid loss of information on the solution structure and as, secondly, other constraints with local variables of the Benders SPs of this set of constraints do not have an RHS unequal to zero. Moreover, in the following, we denote dual variables only for fixing constraints in the sense of procedures (P1) and (P2), as those are the only relevant dual variables for cut generation when using modified constraints for Benders SPs.

Table 4.1 Linking variables pertaining to the different decompositions BO, BO^{dis}, BOEP, BOESP and BOESEP.

Decomposition	Linking Variables
BO	$\dot{E}^{\text{avail}}_{b,u,i}$, $Y^{\text{eff}}_{b,e,i}$
BOEP	$\dot{E}^{\text{avail}}_{b,u,i}$, $X^{\text{mea}}_{b,m,i}$
BOESP	$X^{\text{dev}}_{b,u,i}$, $Y^{\text{eff}}_{b,e,i}$
BOESEP	$X^{\text{dev}}_{b,u,i}$, $X^{\text{mea}}_{b,m,i}$

4.2.1 Building Efficiency Pathways

In the case that constraints pertaining to building efficiency pathways are added to Benders SPs, i.e., in decompositions BOEP and BOESEP, variables $X^{\text{mea}}_{b,m,i}$ are linking variables of the decompositions. Therefore, we apply both procedures (P1) and (P2) to constraints (3.16) – (3.19). Therefore, we replace constraints (3.16) – (3.18) with constraints

$$\sum_{e\in E}\sum_{k=i}^{j} z^{\text{eff}}_{m,e} Y^{\text{eff}}_{b,e,k} \geq (j-i+1)\Big(X^{\text{mea,sp}}_{b,m,i} - \sum_{\substack{\hat{m}\in M:\\ \hat{m}\neq m}} \sum_{k=i+1}^{j} z^{\text{mea}}_{m,\hat{m}} X^{\text{mea,sp}}_{b,\hat{m},k}\Big) \quad \forall b\in B, m\in M, i,j\in I: j\geq i \tag{4.1}$$

$$(i+1)\sum_{j=0}^{i} X^{\text{mea,sp}}_{b,m,j} \geq \sum_{e\in E}\sum_{j=0}^{i} z^{\text{eff}}_{m,e} Y^{\text{eff}}_{b,e,j} \quad \forall b\in B, m\in M: z^{\text{eff}}_{m,e^{\text{exeff}}_b}=0, i\in I \tag{4.2}$$

$$\sum_{e\in E} Y^{\text{eff}}_{b,e,i} = X^{\text{one,eff}}_b \quad \forall b\in B, i\in I \tag{4.3}$$

$$X^{\text{mea,sp}}_{b,m,i} = X^{\text{mea}}_{b,m,i} \quad [\pi^{\text{mea}}_{b,m,i}]\ \forall b\in B, m\in M, i\in I \tag{4.4}$$

$$X^{\text{one,eff}}_b = 1 \quad [\pi^{\text{one,eff}}_b]\ \forall b\in B \tag{4.5}$$

$$X^{\text{one,eff}} \in \mathbb{R}_{\geq 0} \tag{4.6}$$

$$X^{\text{mea,sp}}_{b,m,i} \in \mathbb{R}_{\geq 0} \quad \forall b\in B, m\in M, i\in I. \tag{4.7}$$

We introduce non-negative continuous variables $X^{\text{one,eff}}_b$ and replace constant 1, i.e., the RHS, in constraints (4.3). We then fix the value of $X^{\text{one,eff}}_b$ with constraints (4.5). Note that it is important to define $X^{\text{one,eff}}_b$ for all $b\in B$ to enable decomposability of Benders SPs by buildings $b\in B$. Furthermore, we introduce non-negative continuous variables $X^{\text{mea,sp}}_{b,m,i}$ and use them instead of linking variables $X^{\text{mea}}_{b,m,i}$ in constraints

(4.1) and (4.2). Variables $X_{b,m,i}^{\text{mea,sp}}$ are then fixed to linking variables $X_{b,m,i}^{\text{mea}}$ by means of constraints (4.4). For constraints (4.4) and (4.5), we denote the corresponding dual variables with $\pi_{b,m,i}^{\text{mea}}$ and $\pi_{b}^{\text{one,eff}}$, respectively.

4.2.2 Building Energy System Pathways

In both decompositions BOESP and BOESEP, constraints (3.21) – (3.32) pertaining to building energy system pathways are added to Benders SPs and variables $X_{b,u,i}^{\text{dev}}$ are linking variables. Therefore, we apply both procedures (P1) and (P2) to constraints (3.21) – (3.32). We replace constraints (3.21) – (3.29) with constraints

$$\dot{E}_{b,u,i,j}^{\text{dev}} \leq o_{b,u}^{\text{dev}} X_{b,u,i}^{\text{dev,sp}} \qquad \forall b \in B, u \in U, i \in I, j \in I_{u,i}^{\text{dev}} \tag{4.8}$$

$$\dot{E}_{b,u,i,j}^{\text{dev}} \leq o_{b,u}^{\text{dev}} \Big(X_{b}^{\text{one,dev}} - \sum_{v \in U} z_{u,v}^{\text{dev}} X_{b,v,j}^{\text{dev,sp}} \Big) \qquad \forall b \in B, u \in U \setminus U^{\text{roof}}, i \in I, j \in I_{u,i}^{\text{dev}} \setminus \{i\} \tag{4.9}$$

$$\dot{E}_{b,u,i}^{\text{exdev}} \leq X_{b,u}^{\text{roofdev}} \qquad \forall b \in B, u \in U^{\text{roof}}, i \in I_{b,u}^{\text{exdev}} \tag{4.10}$$

$$\dot{E}_{b,u,i}^{\text{exdev}} \leq o_{b,u}^{\text{exdev}} \Big(X_{b}^{\text{one,dev}} - \sum_{v \in U} z_{u,v}^{\text{dev}} X_{b,v,j}^{\text{dev,sp}} \Big) \qquad \forall b \in B, u \in U \setminus U^{\text{roof}}, i \in I_{b,u}^{\text{exdev}}, j \in \{0, \dots, i\} \tag{4.11}$$

$$\sum_{u \in U_p^{\text{heat}}} \dot{E}_{b,u,i}^{\text{avail}} \geq X_{b,p}^{\text{peak}} \qquad \forall b \in B : \text{th} \notin G_b^{\text{imp}}, p \in P, i \in I \tag{4.12}$$

$$X_{b,i}^{\text{roof,max}} \geq \sum_{u \in U^{\text{roof}}} \sum_{j \in I_{u,i}^{\text{avail}}} \dot{E}_{b,u,j,j}^{\text{dev}} \qquad \forall b \in B, i \in I \tag{4.13}$$

$$X_{b,u,i}^{\text{dev,sp}} = X_{b,u,i}^{\text{dev}} \qquad [\pi_{b,u,i}^{\text{dev}}] \qquad \forall b \in B, u \in U, i \in I \tag{4.14}$$

$$X_{b}^{\text{one,dev}} = 1 \qquad [\pi_{b}^{\text{one,dev}}] \qquad \forall b \in B \tag{4.15}$$

$$X_{b,u}^{\text{roofdev}} = a_{b,u}^{\text{roofdev}} \qquad [\pi_{b,u}^{\text{roofdev}}] \qquad \forall b \in B, u \in U^{\text{roof}} \tag{4.16}$$

$$X_{b,p}^{\text{peak}} = p_{b,p}^{\text{peak}} \qquad [\pi_{b,p}^{\text{peak}}] \qquad \forall b \in B, p \in P \tag{4.17}$$

$$X_{b,i}^{\text{roof,max}} = a_{b,i}^{\text{avail}} \qquad [\pi_{b,i}^{\text{roof,max}}] \qquad \forall b \in B, i \in I \tag{4.18}$$

$$X_{b}^{\text{one,dev}} \in \mathbb{R}_{\geq 0} \qquad \forall b \in B \tag{4.19}$$

$$X_{b,u}^{\text{roof}} \in \mathbb{R}_{\geq 0} \qquad \forall b \in B, u \in U^{\text{roof}} \tag{4.20}$$

$$X_{b,p}^{\text{peak}} \in \mathbb{R}_{\geq 0} \qquad \forall b \in B, p \in P \tag{4.21}$$

$$X_{b,i}^{\text{roof,max}} \in \mathbb{R}_{\geq 0} \qquad \forall b \in B, i \in I \tag{4.22}$$

$$X_{b,u,i}^{\text{dev,sp}} \in \mathbb{R}_{\geq 0} \qquad \forall b \in B, u \in U, i \in I \tag{4.23}$$

Like in the modification for constraints pertaining to building efficiency pathways, we introduce non-negative continuous variables $X_b^{\text{one,dev}}$ and replace constant 1 in constraints (4.9) and (4.11). Variables $X_b^{\text{one,dev}}$ are then fixed with constraints (4.15). Furthermore, we respectively replace constants $a_{b,u}^{\text{roofdev}}$, $p_{b,p}^{\text{peak}}$ and $a_{b,i}^{\text{avail}}$ in constraints (4.10), (4.12) and (4.13) by introducing non-negative continuous variables $X_{b,u}^{\text{roofdev}}$, $X_{b,p}^{\text{peak}}$ and $X_{b,i}^{\text{roof,max}}$ and fixing them to aforementioned parameters by means of constraints (4.16) – (4.18). Next, we introduce non-negative continuous variables $X_{b,u,i}^{\text{dev,sp}}$ and use them instead of linking variables $X_{b,u,i}^{\text{dev}}$ in constraints (4.8), (4.9) and (4.11). Variables $X_{b,u,i}^{\text{dev,sp}}$ are then fixed to the values of linking variables $X_{b,u,i}^{\text{dev}}$ with constraints (4.14). Lastly, we denote the corresponding dual variables of constraints (4.14) – (4.18) with $\pi_{b,u,i}^{\text{dev}}$, $\pi_b^{\text{one,dev}}$, $\pi_{b,u}^{\text{roofdev}}$, $\pi_{b,p}^{\text{peak}}$ and $\pi_{b,i}^{\text{roof,max}}$.

4.2.3 Building Operations

In all decompositions developed here, we use constraints pertaining to building operations in Benders SPs. However, the exact modification of the constraints depends on other constraints and variables not belonging to building operations that are added to the Benders SPs since linking variables change with the chosen model decomposition. Therefore, in a first step, we propose a first modification of constraints that may be modified again, depending on the particular model decomposition. We replace constraints (3.42), (3.41) and (3.46) with constraints

$$\dot{E}_{b,\text{el},i,d,\tau}^{\text{dem}} = X_{b,i,d,\tau}^{\text{el}} \qquad \forall b \in B, i \in I, d \in D, \tau \in T \tag{4.24}$$

$$\dot{E}_{b,\text{th},i,d,\tau}^{\text{dem}} = \sum_{e \in E} p_{b,e,d,\tau}^{\text{heat}} Y_{b,e,i}^{\text{eff}} + X_{b,i,d,\tau}^{\text{dhw}} \qquad \forall b \in B, i \in I, d \in D, \tau \in T \tag{4.25}$$

$$\bar{t}^{\text{dhw}} X_{b,i,d,\tau}^{\text{dhw}} \leq \sum_{u \in U_{\text{dhw}}^{\text{heat}}} \eta_{b,u,\text{th},d,\tau} \dot{E}_{b,u,i,d,\tau}^{\text{op}} \qquad \forall b \in B : \text{th} \notin G_b^{\text{imp}}, i \in I, d \in D, \tau \in T \tag{4.26}$$

$$\bar{t}^{\text{dhw}} X_{b,i,d,\tau}^{\text{dhw}} \leq \sum_{u \in U_{\text{dhw}}^{\text{heat}}} \eta_{b,u,\text{th},d,\tau} \dot{E}_{b,u,i,d,\tau}^{\text{op}} - \dot{E}_{b,\text{th},i,d,\tau}^{\text{imp}} \qquad \forall b \in B : \text{th} \in G_b^{\text{imp}}, i \in I, d \in D, \tau \in T \tag{4.27}$$

$$X^{\text{el}}_{b,i,d,\tau} = p^{\text{el}}_{b,d,\tau} \qquad [\pi^{\text{el}}_{b,i,d,\tau}] \quad \forall b \in B, i \in I, d \in D, \tau \in T \tag{4.28}$$

$$X^{\text{dhw}}_{b,i,d,\tau} = p^{\text{dhw}}_{b,d,\tau} \qquad [\pi^{\text{dhw}}_{b,i,d,\tau}] \quad \forall b \in B, i \in I, d \in D, \tau \in T \tag{4.29}$$

$$X^{\text{el}}_{b,i,d,\tau} \in \mathbb{R}_{\geq 0} \qquad \forall b \in B, i \in I, d \in D, \tau \in T \tag{4.30}$$

$$X^{\text{dhw}}_{b,i,d,\tau} \in \mathbb{R}_{\geq 0} \qquad \forall b \in B, i \in I, d \in D, \tau \in T \tag{4.31}$$

We introduce non-negative continuous variables $X^{\text{el}}_{b,i,d,\tau}$ and $X^{\text{dhw}}_{b,i,d,\tau}$ to replace the RHS in constraints (4.24) – (4.27), respectively. Constraints (4.28) and (4.29) fix the values of variables $X^{\text{el}}_{b,i,d,\tau}$ and $X^{\text{dhw}}_{b,i,d,\tau}$ to the respective RHS. We denote dual variables of constraints (4.28) and (4.29) with $\pi^{\text{el}}_{b,i,d,\tau}$ and $\pi^{\text{dhw}}_{b,i,d,\tau}$, respectively. Note that we only include indices $i \in I$ for variables $X^{\text{el}}_{b,i,d,\tau}$ and $X^{\text{dhw}}_{b,i,d,\tau}$ in constraints (4.28) – (4.31) if we separate Benders cuts disaggregated by years $i \in I$ with model decomposition BO but only state constraints for this case to simplify notation.

In decompositions BO and BOESP, variables $Y^{\text{eff}}_{b,e,i}$ represent linking variables. We therefore replace constraints (4.25) with constraints

$$\dot{E}^{\text{dem}}_{b,\text{th},i,d,\tau} = \sum_{e \in E} p^{\text{heat}}_{b,e,d,\tau} X^{\text{eff,sp}}_{b,e,i} + X^{\text{dhw}}_{b,i,d,\tau} \qquad \forall b \in B, i \in I, d \in D, \tau \in T \tag{4.32}$$

$$X^{\text{eff,sp}}_{b,e,i} = Y^{\text{eff}}_{b,e,i} \qquad [\pi^{\text{eff}}_{b,e,i}] \quad \forall b \in B, e \in E, i \in I \tag{4.33}$$

$$X^{\text{eff,sp}}_{b,e,i} \in \mathbb{R}_{\geq 0} \qquad \forall b \in B, e \in E, i \in I. \tag{4.34}$$

Here, we introduce non-negative continuous variables $X^{\text{eff,sp}}_{b,e,i}$ to replace linking variables $Y^{\text{eff}}_{b,e,i}$ in constraints (4.32). The values of variables $X^{\text{eff,sp}}_{b,e,i}$ are then fixed by means of constraints (4.33). We denote dual variables that correspond to constraints (4.33) with $\pi^{\text{eff}}_{b,e,i}$.

In decompositions BO and BOEP, variables $\dot{E}^{\text{avail}}_{b,u,i}$ are linking variables. In those decompositions, we replace constraints (3.35) – (3.39) with constraints

$$\dot{E}^{\text{op}}_{b,u,i,d,\tau} \leq \min_{f \in H^{\text{out}}_u} \{\eta_{b,u,f,d,\tau}\}^{-1} X^{\text{avail,sp}}_{b,u,i} \qquad \forall b \in B, u \in U \setminus (U^{\text{sto}} \cup U^{\text{roof}}), \; i \in I, d \in D, \tau \in T \tag{4.35}$$

$$\dot{E}^{\text{op}}_{b,u,i,d,\tau} \leq g_{b,d,\tau} X^{\text{avail,sp}}_{b,u,i} \qquad \forall b \in B, u \in U^{\text{roof}}, i \in I, d \in D, \tau \in T \tag{4.36}$$

$$SOC_{b,u,i,d,\tau} \leq dod_u X^{\text{avail,sp}}_{b,u,i} \qquad \forall b \in B, u \in U^{\text{sto}}, i \in I, d \in D, \tau \in T \tag{4.37}$$

$$CH_{b,u,i,d,\tau} \leq \frac{dod_u}{\eta^{\text{ch}}_u \Delta\tau^{\min}_u} X^{\text{avail,sp}}_{b,u,i} \qquad \forall b \in B, u \in U^{\text{sto}}, i \in I, d \in D, \tau \in T \tag{4.38}$$

$$DCH_{b,u,i,d,\tau} \leq \frac{dod_u}{\Delta\tau^{\min}_u} X^{\text{avail,sp}}_{b,u,i} \qquad \forall b \in B, u \in U^{\text{sto}}, i \in I, d \in D, \tau \in T \tag{4.39}$$

$$X^{\text{avail,sp}}_{b,u,i} = \dot{E}^{\text{avail}}_{b,u,i} \qquad [\pi^{\text{avail}}_{b,u,i}] \qquad \forall b \in B, u \in U, i \in I \tag{4.40}$$

$$X^{\text{avail,sp}}_{b,u,i} \in \mathbb{R}_{\geq 0} \qquad \forall b \in B, u \in U, i \in I. \tag{4.41}$$

We introduce non-negative continuous variables $X^{\text{avail,sp}}_{b,u,i}$ and constraints (4.35) – (4.39) to replace linking variables in constraints (3.35) – (3.39). Constraints (4.40) fix the values of variables $X^{\text{avail,sp}}_{b,u,i}$ to linking variables $\dot{E}^{\text{avail}}_{b,u,i}$. Dual variables that correspond to constraints (4.40) are denoted with $\pi^{\text{avail}}_{b,u,i}$.

4.3 Model Decompositions

In this section, we present the different model decompositions BO, BOEP, BOESP, and BOESEP. For a better overview, we refer to the constraints of the original MILP model presented in Chap. 3 or their modifications introduced in Sect. 4.2. In Benders SPs, we show constraints pertaining to the standard decomposition as well as constraints pertaining to previously shown modifications. Moreover, we assume linking variables used in respective Benders SPs to be fixed to the values of a Benders RMP solution. Furthermore, we assume the integrality condition on all binary linking variables to be relaxed in Benders SPs. Lastly, we introduce a technique to strengthen Benders optimality cuts for decompositions BO and BOESP as well as a set of valid inequalities to be used in Benders RMPs of decompositions BOEP, BOESP, and BOESEP.

4.3.1 Building Operations (BO)

Benders SPs

In this first decomposition, we move constraints corresponding to building operations to Benders SPs. For a fixed building $b \in B$ and year $i \in I$, the operations in b and i are decomposable for all typical days $d \in D$. However, generated cuts for the Benders RMP do not contain variables defined in the domain of typical days D and thus can be aggregated for every year. Since the creation of an excessive number of subprocesses needed for the parallelization of Benders cut separation may create computational overhead, we decide to disaggregate Benders SPs only by buildings and years. Therefore, we consider Benders SPs $k \in \{0, \ldots, n^{\text{sp}} - 1\}$ with which we partition sets B and I into non-empty and disjoint sets $B_k^{\text{sp}} \subseteq B$ and $I_k^{\text{sp}} \subseteq I$. Then, we define Benders SPs $(\text{SP}_k^{\text{BO}})$ for decomposition BO with

$$(\text{SP}_k^{\text{BO}}) \quad \min \sum_{b \in B_k^{\text{sp}}} \sum_{i \in I_k^{\text{sp}}} O_{b,i}^{\text{sp}} \tag{4.42}$$

$$\text{s.t.} \quad O_{b,i}^{\text{sp}} = \phi\, C_{b,i}^{\text{en,sp}} + (1 - \phi)\, \omega\, TGW_{b,i}^{\text{op}} \qquad \forall b \in B_k^{\text{sp}}, i \in I_k^{\text{sp}} \tag{4.43}$$

$$\text{(3.19) and (3.32)} \qquad \forall b \in B_k^{\text{sp}}, i \in I_k^{\text{sp}} \tag{4.44}$$

$$\text{(3.35) – (3.54)} \ \underline{\text{or (3.40), (3.43) – (3.45),}}$$
$$\underline{\text{(3.48) – (3.54), (4.24) and (4.26) – (4.41)}} \qquad \forall b \in B_k^{\text{sp}}, i \in I_k^{\text{sp}} \tag{4.45}$$

$$\text{(3.67), (3.69), (3.83) and (3.84)} \qquad \forall b \in B_k^{\text{sp}}, i \in I_k^{\text{sp}} \tag{4.46}$$

$$O_{b,i}^{\text{sp}} \in \mathbb{R} \qquad \forall b \in B_k^{\text{sp}}, i \in I_k^{\text{sp}} \tag{4.47}$$

where we assume linking variables $\dot{E}_{b,u,i}^{\text{avail}}$ and $Y_{b,e,i}^{\text{eff}}$ to be fixed to values of a solution of the Benders RMP and constraints in (4.44) – (4.46) to be set up for all $b \in B : b \in B_k^{\text{sp}}$ and $i \in I : i \in I_k^{\text{sp}}$. Continuous variables $O_{b,i}^{\text{sp}}$ defined in (4.47) correspond to objective values with respect to (4.43) for optimization of building operations for building b in year i. Constraints in (4.44) and (4.45) pertain to linking variables declarations as well as constraints of building operations and cost and emission calculation, respectively. Underlined constraints denote optional modified constraints as presented in Sect. 4.2.

Benders RMP

The Benders RMP (RMP^{BO}) for this decomposition reads

$$
(\text{RMP}^{\text{BO}}) \quad
\begin{aligned}
\min \quad & \phi\left(C^{\text{inv}} + C^{\text{om}}\right) + (1-\phi)\,\omega\, TGW^{\text{mea}} + \sum_{b\in B}\sum_{i\in I} \eta_{b,i} && (4.48)\\
\text{s.t.} \quad & (3.5) - (3.11) \text{ and } (3.13) - (3.15) && (4.49)\\
& (3.16) - (3.19) && (4.50)\\
& (3.21) - (3.32) && (4.51)\\
& (3.57) - (3.65),\ (3.70),\ (3.73) - (3.81) \text{ and } (3.85) && (4.52)\\
& \text{Benders cuts} && (4.53)\\
& \eta_{b,i} \in \mathbb{R} \qquad \forall b \in B, i \in I, && (4.54)
\end{aligned}
$$

where $\eta_{b,i}$ are continuous variables that are used for the construction of Benders cuts in (4.53). Like Maher (2021), we henceforth call variables in the form of $\eta_{b,i}$ *underestimators* of the objective values of the Benders SPs. Benders cuts are constructed for all $(b, i) \in B \times I$ and only consist of Benders optimality cuts because of constraints (3.28) that ensure that peak demands $p^{\text{peak}}_{b,p}$ for $p \in P$ are covered in every building $b \in B$ and year $i \in I$. Since $p^{\text{peak}}_{b,\text{th}} \geq \max_{e\in E, d\in D, \tau\in T}\{p^{\text{dhw}}_{b,d,\tau} + p^{\text{heat}}_{b,e,d,\tau}\}$ as well as $p^{\text{peak}}_{b,\text{dhw}} \geq \tilde{t}^{\text{dhw}} \max_{d\in D,\tau\in T}\{p^{\text{dhw}}_{b,d,\tau}\}$ for all $b \in B$, cf. Sect. 3.1.5, demands in constraints (3.41) and (3.46) can be covered for all $d \in D$ and $\tau \in T$. Furthermore, as electrical energy can be freely exchanged with the electricity grid, every fixed solution in linking variables $Y^{\text{eff}}_{b,e,i}$ and $\dot{E}^{\text{avail}}_{b,u,i}$ thus leads to feasible (SP^{BO}_k) for all $k \in \{0, \ldots, n^{\text{sp}} - 1\}$. The set of Benders cuts in (4.53) therefore only contains Benders optimality cuts. In the following, we present Benders optimality cuts for this decomposition based on the modified version of the constraints in (SP^{BO}_k). Benders cuts for the unmodified constraints in (SP^{BO}_k) or for other decompositions BOEP, BOESP, and BOESEP are generated analogously.

Benders Cuts

Let $\left(\overline{\mathbf{Y}^{\text{eff}}_{b,i}}, \overline{\dot{\mathbf{E}}^{\text{avail}}_{b,i}}, \overline{\eta_{b,i}}\right)$ be the values of variables $Y^{\text{eff}}_{b,e,i}$, $\dot{E}^{\text{avail}}_{b,u,i}$ and $\eta_{b,i}$ in a solution of (RMP^{BO}) corresponding to $b \in B^{\text{sp}}_k$, $i \in I^{\text{sp}}_k$ for $k \in \{0, \ldots, n^{\text{sp}} - 1\}$. We then fix linking variables $\mathbf{Y}^{\text{eff}}_{b,i}$ and $\dot{\mathbf{E}}^{\text{avail}}_{b,i}$ in each (SP^{BO}_k) to the corresponding solution of (RMP^{BO}), solve the resulting LP models and retrieve objective values $\overline{O^{\text{sp}}_{b,i}}$ and dual solutions $\left(\overline{\pi^{\text{el}}_{b,i,d,\tau}}, \overline{\pi^{\text{dhw}}_{b,i,d,\tau}}, \overline{\pi^{\text{eff}}_{b,e,i}}, \overline{\pi^{\text{avail}}_{b,u,i}}\right)$, corresponding to fixing constraints (4.28), (4.29), (4.33) and (4.40). Now, for all $b \in B, i \in I : \overline{O^{\text{sp}}_{b,i}} > \overline{\eta_{b,i}}$, we add Benders optimality cuts

$$\eta_{b,i} \geq \sum_{d \in D} \sum_{\tau \in T} \left(\overline{\pi^{\text{el}}_{b,i,d,\tau}} p^{\text{el}}_{b,d,\tau} + \overline{\pi^{\text{dhw}}_{b,d,\tau}} p^{\text{dhw}}_{b,i,d,\tau} \right) - \sum_{e \in E} \overline{\pi^{\text{eff}}_{b,e,i}} Y^{\text{eff}}_{b,e,i} - \sum_{u \in U} \overline{\pi^{\text{avail}}_{b,u,i}} \dot{E}^{\text{avail}}_{b,u,i} \tag{4.55}$$

based on inequality (2.27) to the set of Benders cuts in (4.53). A solution of (RMP) is feasible for the original model if $\overline{O^{\text{sp}}_{b,i}} \leq \overline{\eta_{b,i}}\ \forall b \in B, i \in I$. In the following we will introduce a technique to strengthen Benders optimality cuts (4.55) for decompositions BO and BOESP.

Strengthening of Benders Cuts

Benders optimality cuts in (4.55) are outer approximations of $O^{\text{sp}}_{b,i}$ in (4.43) for given solutions of (RMP^{BO}) that were used to separate the respective Benders optimality cuts. As explained in Sect. 3.3, in any feasible solution, exactly one building efficiency state $e \in E$ is taken in every year $i \in I$ and building $b \in B$, i.e, it is $Y^{\text{eff}}_{b,e,i} = 1$ for exactly one $e \in E$ for fixed $i \in I$ and $b \in B$ and $Y^{\text{eff}}_{b,\hat{e},i} = 0\ \forall \hat{e} \in E : \hat{e} \neq e$. Based on this property, we observe that under certain parametric conditions of problem instances, we can strengthen Benders optimality cuts (4.55) by relating objective values $O^{\text{sp}}_{b,i}$ in (SP$^{\text{BO}}_k$) with variables $Y^{\text{eff}}_{b,e,i}$. In the following, we will show how to strengthen Benders optimality cuts (4.55) and provide sufficient conditions for the instances for the strengthening to be applied.

Let $S^{\text{RMP,BO}}$ as defined in Sect. 2.2 be the feasible region of (RMP^{BO}) with no Benders cuts separated. Henceforth, we use $(\cdot) \in S^{\text{RMP,BO}}$ to denote partial solutions as tuples of vectors that are simultaneously part of feasible solutions of (RMP^{BO}). Furthermore, let $O^{\text{sp}*}_{b,i}\left(\mathbf{Y}^{\text{eff}}_{b,i}, \dot{\mathbf{E}}^{\text{avail}}_{b,i}\right)$ given by (4.43) denote the objective value belonging to an optimal solution of (SP$^{\text{BO}}_k$) where $\left(\mathbf{Y}^{\text{eff}}_{b,i}, \dot{\mathbf{E}}^{\text{avail}}_{b,i}\right) \in S^{\text{RMP,BO}}$. We can strengthen Benders optimality cuts (4.55) if it holds for for all $\left(\mathbf{Y}^{\text{eff}}_{b,i}\right), \left(\mathbf{Y}^{\text{eff}\prime}_{b,i}\right) \in S^{\text{RMP,BO}}$ that either

$$O^{\text{sp}*}_{b,i}\left(\mathbf{Y}^{\text{eff}\prime}_{b,i}, \dot{\mathbf{E}}^{\text{avail}}_{b,i}\right) > O^{\text{sp}*}_{b,i}\left(\mathbf{Y}^{\text{eff}}_{b,i}, \dot{\mathbf{E}}^{\text{avail}}_{b,i}\right) \quad \forall \left(\mathbf{Y}^{\text{eff}\prime}_{b,i}, \dot{\mathbf{E}}^{\text{avail}}_{b,i}\right), \left(\mathbf{Y}^{\text{eff}}_{b,i}, \dot{\mathbf{E}}^{\text{avail}}_{b,i}\right) \in S^{\text{RMP,BO}} \tag{4.56}$$

or (exclusively)

$$O^{\text{sp}*}_{b,i}\left(\mathbf{Y}^{\text{eff}\prime}_{b,i}, \dot{\mathbf{E}}^{\text{avail}}_{b,i}\right) \leq O^{\text{sp}*}_{b,i}\left(\mathbf{Y}^{\text{eff}}_{b,i}, \dot{\mathbf{E}}^{\text{avail}}_{b,i}\right) \forall \left(\mathbf{Y}^{\text{eff}\prime}_{b,i}, \dot{\mathbf{E}}^{\text{avail}}_{b,i}\right), \left(\mathbf{Y}^{\text{eff}}_{b,i}, \dot{\mathbf{E}}^{\text{avail}}_{b,i}\right) \in S^{\text{RMP,BO}}. \tag{4.57}$$

If the aforementioned property is satisfied, this implies that for a fixed $b \in B$ and $i \in I$, optimal objective values $O^{\text{sp}*}_{b,i}$ are valid upper or lower bounds on each other for two taken building efficiency states $e, \hat{e} \in E$, irrespective of the building energy system design $\dot{\mathbf{E}}^{\mathrm{avail}}_{b,i}$. We can therefore define sets

$$W_{b,e,i} = \Big\{ \hat{e} \in E : O^{\mathrm{sp}*}_{b,i}\left(\mathbf{Y}^{\mathrm{eff}\,\prime}_{b,i}, \dot{\mathbf{E}}^{\mathrm{avail}}_{b,i}\right) \leq O^{\mathrm{sp}*}_{b,i}\left(\mathbf{Y}^{\mathrm{eff}}_{b,i}, \dot{\mathbf{E}}^{\mathrm{avail}}_{b,i}\right) \\ \forall \left(\mathbf{Y}^{\mathrm{eff}\,\prime}_{b,i}, \dot{\mathbf{E}}^{\mathrm{avail}}_{b,i}\right), \left(\mathbf{Y}^{\mathrm{eff}}_{b,i}, \dot{\mathbf{E}}^{\mathrm{avail}}_{b,i}\right) \in S^{\mathrm{RMP,BO}} : \\ Y^{\mathrm{eff}\,\prime}_{b,\hat{e},i} = 1, Y^{\mathrm{eff}}_{b,e,i} = 1 \Big\} \qquad \forall b \in B, e \in E, i \in I. \qquad (4.58)$$

Set $W_{b,e,i}$, therefore, contains the building efficiencies $\hat{e} \in E$ for which the Benders SP ($\mathrm{SP}^{\mathrm{BO}}_k$) corresponding to building $b \in B$ and year $i \in I$ always yields an optimal objective value smaller or equal to the one obtained with $e \in E$. Then, instead of regular Benders optimality cuts (4.55), we can use strengthened Benders optimality cuts

$$\eta_{b,i} \geq \underbrace{\sum_{d \in D} \sum_{\tau \in T} \left(\overline{\pi^{\mathrm{el}}_{b,i,d,\tau}} p^{\mathrm{el}}_{b,d,\tau} + \overline{\pi^{\mathrm{dhw}}_{b,i,d,\tau}} p^{\mathrm{dhw}}_{b,d,\tau} \right)}_{=\overline{\alpha}_{b,i}} - \sum_{\hat{e} \in E} \min_{e \in W_{b,\hat{e},i}} \left\{ \overline{\pi^{\mathrm{eff}}_{b,e,i}} \right\} Y^{\mathrm{eff}}_{b,e,i} \\ - \sum_{u \in U} \overline{\pi^{\mathrm{avail}}_{b,u,i}} \dot{E}^{\mathrm{avail}}_{b,u,i}. \qquad (4.59)$$

for a given building $b \in B$ and year $i \in I$. The inequality in (4.59) is valid for the original MILP model since for a given $b \in B$ and $i \in I$ we have for every $\left(\widetilde{\mathbf{Y}^{\mathrm{eff}}_{b,i}}, \widetilde{\dot{\mathbf{E}}^{\mathrm{avail}}_{b,i}}\right) \in S^{\mathrm{RMP,BO}}$ where $\widetilde{Y^{\mathrm{eff}}_{b,\hat{e},i}} = 1$ that

$$\begin{aligned} O^{\mathrm{sp}*}_{b,i}\left(\widetilde{\mathbf{Y}^{\mathrm{eff}}_{b,i}}, \widetilde{\dot{\mathbf{E}}^{\mathrm{avail}}_{b,i}}\right) &\overset{(4.58)}{\geq} \max_{e \in W_{b,\hat{e},i}} \left\{ \overline{\alpha}_{b,i} - \overline{\pi^{\mathrm{eff}}_{b,e,i}} - \sum_{u \in U} \overline{\pi^{\mathrm{avail}}_{b,u,i}} \widetilde{\dot{E}^{\mathrm{avail}}_{b,u,i}} \right\} \\ &= \overline{\alpha}_{b,i} - \min_{e \in W_{b,\hat{e},i}} \left\{ \overline{\pi^{\mathrm{eff}}_{b,e,i}} \right\} - \sum_{u \in U} \overline{\pi^{\mathrm{avail}}_{b,u,i}} \widetilde{\dot{E}^{\mathrm{avail}}_{b,u,i}} \\ &\geq \overline{\alpha}_{b,i} - \overline{\pi^{\mathrm{eff}}_{b,\hat{e},i}} - \sum_{u \in U} \overline{\pi^{\mathrm{avail}}_{b,u,i}} \widetilde{\dot{E}^{\mathrm{avail}}_{b,u,i}}, \end{aligned}$$

i.e., for all feasible solutions $\left(\widetilde{\mathbf{Y}^{\mathrm{eff}}_{b,i}}, \widetilde{\dot{\mathbf{E}}^{\mathrm{avail}}_{b,i}}\right) \in S^{\mathrm{RMP,BO}}$, the strengthened Benders cuts (4.59) underestimate the respective optimal objective value $O^{\mathrm{sp}*}_{b,i}\left(\widetilde{\mathbf{Y}^{\mathrm{eff}}_{b,i}}, \widetilde{\dot{\mathbf{E}}^{\mathrm{avail}}_{b,i}}\right)$ for all fixed $b \in B$ an $i \in I$ and, therefore, no feasible integer solution of the original MILP model is cut off. Benders optimality cuts in (4.59) being stronger than the ones in (4.55) follows from the last inequality. The presented strengthening technique can be used for decompositions BO and BOESP where linking variables $Y^{\mathrm{eff}}_{b,e,i}$ are in Benders SPs.

In the following, we informally present sufficient conditions for an instance parametrization for the aforementioned property to hold such that the strengthening of Benders optimality cuts can be applied. Here, we only consider the case of $\phi = 1$, i.e., the sole minimization of $-NPV$ in (3.55). A problem instance with $\phi = 1$ has the aforementioned property if the following conditions are fulfilled:

(C1) It holds that

$$\eta_{b,\mathrm{CHP,el},d,\tau} s_{i,\mathrm{el}} < c_{i,\mathrm{gas}} \quad \forall b \in B, i \in I, d \in D, \tau \in T. \tag{4.60}$$

(C2) For all buildings $b \in B$ and all $e, \hat{e} \in E$ it holds that either

$$p^{\mathrm{th}}_{b,e,d,\tau} \geq p^{\mathrm{th}}_{b,\hat{e},d,\tau} \qquad \forall d \in D, \tau \in T \tag{4.61}$$

or (exclusively)

$$p^{\mathrm{th}}_{b,e,d,\tau} < p^{\mathrm{th}}_{b,\hat{e},d,\tau} \qquad \forall d \in D, \tau \in T. \tag{4.62}$$

Condition (C1) implies that all devices $u \in U : \mathrm{th} \in H^{\mathrm{out}}_u$, including CHP, have non-negative marginal costs related to the production of thermal energy. To see why the property holds with the additional condition (C2), consider an optimal energy system operation for a given $b \in B$ and $i \in I$ with a current building efficiency state $\hat{e} \in E$. If $p^{\mathrm{th}}_{b,e,d,\tau} \geq p^{\mathrm{th}}_{b,\hat{e},d,\tau} \; \forall d \in D, \tau \in T$, the non-negative increase in thermal energy demand for all $d \in D$ and $\tau \in T$ entails a non-negative increase in costs. Since this holds for all $d \in D$ and $\tau \in T$, the total associated costs connected with building efficiency $\hat{e}$ cannot be lower than with e. The opposite case follows from the same observation.

4.3.2 Building Operations and Efficiency Pathways (BOEP)

Benders SPs

In the decomposition BOEP, we include constraints and variables pertaining to building efficiency pathways and building operations in Benders SPs. Constraints (3.16) and (3.17) in building efficiency pathways link consecutive years, which is why it is not possible to decompose Benders SPs in terms of years $i \in I$. Therefore, we introduce Benders SP ($\text{SP}_k^{\text{BOEP}}$) $\forall k \in \{0, \ldots, n^{\text{sp}} - 1\}$ to partition the set B into non-empty and disjoint sets $B_k^{\text{sp}} \subseteq B$. The Benders SP ($\text{SP}_k^{\text{BOEP}}$) is then given by

$$\min \sum_{b \in B_k^{\text{sp}}} O_b^{\text{sp}} \tag{4.63}$$

$$(\text{SP}_k^{\text{BOEP}}) \quad \text{s.t.} \quad O_b^{\text{sp}} = \sum_{i \in I} \left(\phi\, C_{b,i}^{\text{en,sp}} + (1 - \phi)\, \omega\, TGW_{b,i}^{\text{op}} \right) \qquad \forall b \in B_k^{\text{sp}} \tag{4.64}$$

$$(3.14) \text{ and } (3.32) \qquad \forall b \in B_k^{\text{sp}} \tag{4.65}$$

$$(3.16) - (3.19) \ \underline{\text{or } (3.19) \text{ and } (4.1) - (4.7)} \qquad \forall b \in B_k^{\text{sp}} \tag{4.66}$$

$$(3.35) - (3.54) \ \underline{\text{or } (3.35) - (3.40),\ (3.43) - (3.45),}$$

$$\underline{(3.48) - (3.54),\ (4.24) - (4.31) \text{ and } (4.35) - (4.41)} \qquad \forall b \in B_k^{\text{sp}} \tag{4.67}$$

$$(3.67),\ (3.69),\ (3.83) \text{ and } (3.84) \qquad \forall b \in B_k^{\text{sp}} \tag{4.68}$$

$$O_b^{\text{sp}} \in \mathbb{R} \qquad \forall b \in B_k^{\text{sp}}, \tag{4.69}$$

where we assume linking variables $\dot{E}_{b,u,i}^{\text{avail}}$ and $X_{b,m,i}^{\text{mea}}$ to be fixed to values of a solution of the Benders RMP and constraints in (4.65) – (4.68) to be set up for $b \in B : b \in B_k^{\text{sp}}$. Continuous variables O_b^{sp} defined in (4.69) measure the objective function for each $b \in B_k^{\text{sp}}$ by means of constraints (4.64). The objective expressed in constraints (4.64) is equivalent to the year-wise aggregated objective function of decomposition BO in (4.43) since variables $Y_{b,e,i}^{\text{eff}}$ pertaining to building efficiency pathways do not have objective coefficients unequal to zero in (3.55) of the original MILP model. Constraints in (4.65) – (4.68) pertain to the declaration of linking variables as well as constraints of building efficiency pathways, building operations and cost and emission calculation, respectively. Underlined constraints denote optional modified constraints as presented in Sect. 4.2.

Valid Inequalities

With constraints (3.16) – (3.19) being in Benders SPs, in solutions of the Benders RMP, more than one efficiency measure $m \in M$ of the same efficiency measure class $c \in C^{\text{mea}}$ might be realized for the same building $b \in B$ and year $i \in I$. However, by construction of the set of building efficiencies E, for any $c \in C^{\text{mea}}$ there exists no $e \in E : \sum_{m \in M_c} z_{m,e}^{\text{eff}} > 1$ and constraints (3.16) – (3.19) therefore ensure that at most one efficiency measure $m \in M_c$ for any $c \in C^{\text{mea}}$ is realized in building $b \in B$ in the same year $i \in I$. For this reason, the aforementioned Benders RMP solutions would lead to infeasibility of ($\text{SP}_k^{\text{BOEP}}$). For this reason, we introduce valid inequalities

$$\sum_{m \in M_c} X^{\text{mea}}_{b,m,i} \leq 1 \qquad \forall b \in B, c \in C^{\text{mea}}, i \in I \tag{4.70}$$

that avoid that Benders SPs ($\text{SP}^{\text{BOEP}}_k$) are infeasible because of aforementioned solutions of (RMP^{BOEP}). Therefore, constraints (4.70) retain restrictions in variables $X^{\text{mea}}_{b,m,i}$ originally imposed by constraints (3.16) – (3.19) and thus ensure that problems ($\text{SP}^{\text{BOEP}}_k$) are always feasible. Consequently, the set of Benders cuts in (4.76) only consist of optimality cuts.

Benders RMP

The corresponding Benders RMP (RMP^{BOEP}) is given by

$$(\text{RMP}^{\text{BOEP}}) \quad \begin{aligned} \min \quad & \phi \left(C^{\text{inv}} + C^{\text{om}}\right) + (1 - \phi)\, \omega\, TGW^{\text{mea}} + \sum_{b \in B} \eta_b && (4.71) \\ \text{s.t.} \quad & (3.5) - (3.11) \text{ and } (3.13) - (3.15) && (4.72) \\ & (3.21) - (3.32) && (4.73) \\ & (3.57) - (3.65) \text{ and } (3.70), && \\ & (3.73) - (3.81) \text{ and } (3.85) && (4.74) \\ & (4.70) && (4.75) \\ & \text{Benders cuts} && (4.76) \\ & \eta_b \in \mathbb{R} \qquad \forall b \in B, && (4.77) \end{aligned}$$

where we introduce continuous variables η_b that are used for the construction of Benders cuts in (4.76). Constraints in (4.72) – (4.74) are constraints and variable declarations pertaining to resource allocation at building stock level, building energy system pathways and cost and emission calculation, respectively. Lastly, constraints (4.75) are valid inequalities that guarantee that Benders SPs ($\text{SP}^{\text{BOEP}}_k$) remain feasible for all solutions of (RMP^{BOEP}). Therefore, the set of Benders cuts in (4.76) only contains Benders optimality cuts.

4.3.3 Building Operations and Energy System Pathways (BOESP)

Benders SPs

In decomposition BOESP, we include constraints and variable declarations belonging to building energy system pathways and building operations in Benders SPs. Similarly to decomposition BOEP, constraints pertaining to building energy system pathways in decomposition BOESP are coupled between years $i \in I$. Therefore, we also introduce a partition of set B into non-empty disjoint sets $B_k^{\text{sp}} \subseteq B$ with $k \in \{0, \ldots, n^{\text{sp}} - 1\}$. The Benders SP ($\text{SP}_k^{\text{BOESP}}$) is then defined by

$$\min \sum_{b \in B_k^{\text{sp}}} O_b^{\text{sp}} \tag{4.78}$$

$(\text{SP}_k^{\text{BOESP}})$

$$\text{s.t.} \quad O_b^{\text{sp}} = \sum_{i \in I} \Big(\phi \, (C_{b,i}^{\text{inv,dev,var}} + C_{b,i}^{\text{om,dev,var}} + C_{b,i}^{\text{en,sp}}) + (1-\phi)\, \omega \, TGW_{b,i}^{\text{op}} \Big) \quad \forall b \in B_k^{\text{sp}} \tag{4.79}$$

$$\text{(3.13) and (3.19)} \quad \forall b \in B_k^{\text{sp}} \tag{4.80}$$

$$\text{(3.21)} - \text{(3.32)} \;\; \underline{\text{or (3.23), (3.26) and (3.27), (3.30)} - \text{(3.32) and (4.8)} - \text{(4.23)}} \quad \forall b \in B_k^{\text{sp}} \tag{4.81}$$

$$\text{(3.35)} - \text{(3.54)} \;\; \underline{\text{or (3.40), (3.43)} - \text{(3.45), (3.48)} - \text{(3.54), (4.24) and (4.26)} - \text{(4.34)}} \quad \forall b \in B_k^{\text{sp}} \tag{4.82}$$

$$\text{(3.59), (3.65), (3.67) and (3.69), (3.75), (3.81), (3.83) and (3.84)} \quad \forall b \in B_k^{\text{sp}} \tag{4.83}$$

$$O_b^{\text{sp}} \in \mathbb{R} \quad \forall b \in B_k^{\text{sp}}, \tag{4.84}$$

where we assume linking variables $X_{b,u,i}^{\text{dev}}$ and $Y_{b,e,i}^{\text{eff}}$ to be fixed to values of a solution of the Benders RMP and constraints in (4.80) – (4.83) to be set up for $b \in B : b \in B_k^{\text{sp}}$. Continuous variables O_b^{sp} defined in (4.84) measure the building-wise objective function through constraints (4.79). The objective in ($\text{SP}_k^{\text{BOESP}}$) differs from the objective function in (SP_k^{BO}) and ($\text{SP}_k^{\text{BOEP}}$) since variables $\dot{E}_{b,u,i}^{\text{avail}}$ are now part of Benders SPs and corresponding objective coefficients are now part of ($\text{SP}_k^{\text{BOESP}}$). Constraints in (4.80) – (4.83) pertain to the declaration of linking variables as well as the constraints of building energy system pathways, building operations and cost and emission calculation, respectively. Underlined constraints denote optional modified constraints as presented in Sect. 4.2.

Valid Inequalities

Due to constraints (3.21) – (3.32) being in Benders SPs being in ($\text{SP}_k^{\text{BOESP}}$), restrictions on the solution structure imposed by those constraints in ($\text{SP}_k^{\text{BOESP}}$) on variables $\dot{E}_{b,u,i}^{\text{avail}}$ is not imposed on variables $X_{b,u,i}^{\text{dev}}$ in ($\text{RMP}^{\text{BOESP}}$). Therefore, solutions of ($\text{RMP}^{\text{BOESP}}$) might lead to infeasibilities of ($\text{SP}_k^{\text{BOESP}}$) because installed devices $u \in U_p^{\text{heat}}$ might not be able to cover peak demand $p_{b,p}^{\text{peak}}$ for $p \in P$. Valid inequalities can retain some of the restrictions originally imposed by constraints that are now included in Benders SPs. To define valid inequalities for this decomposition, we first define sets and

$$I^{\text{cov}}_{b,p} = \Big\{ i \in I : \sum_{\substack{v \in U^{\text{heat}}_p : \\ l^{\text{exdev}}_{b,v} \geq i+1}} o^{\text{exdev}}_{b,v} < p^{\text{peak}}_{b,p} \Big\} \qquad \forall b \in B, p \in P \tag{4.85}$$

and

$$I^{\text{cov,inst}}_{b,p,u} = \Big\{ i \in I \setminus I^{\text{cov}}_{b,p} : \sum_{\substack{v \in U^{\text{heat}}_p : \\ l^{\text{exdev}}_{b,u} \geq i+1}} (1 - z^{\text{dev}}_{u,v})\, o^{\text{exdev}}_{b,v} < p^{\text{peak}}_{b,p} \Big\} \quad \forall b \in B, p \in P, u \in U^{\text{heat}}_p. \tag{4.86}$$

Sets $I^{\text{cov}}_{b,p}$ in (4.85) contain years $i \in I$ for building $b \in B$ for which existing devices are not able to cover peak demand $p^{\text{peak}}_{b,p}$ for $p \in P$. Similarly, sets $I^{\text{cov,inst}}_{b,p,v}$ contain years $i \in I$ for $b \in B$ in which existing devices would be able to cover peak demand $p^{\text{peak}}_{b,p}$ for $p \in P$ but the installation of device u before or in the regarded year leads to existing devices not being to cover $p^{\text{peak}}_{b,p}$. With sets $I^{\text{cov}}_{b,p}$ and $I^{\text{cov,inst}}_{b,p,u}$ given, we formulate the valid inequalities

$$\sum_{u \in U^{\text{heat}}_p} \sum_{j \in I^{\text{avail}}_{u,i}} X^{\text{dev}}_{b,u,j} \geq 1 \qquad \forall b \in B : \text{th} \notin G^{\text{imp}}_b, p \in P, i \in I^{\text{cov}}_{b,p} \tag{4.87}$$

$$X^{\text{dev}}_{b,u,h} + X^{\text{dev}}_{b,v,j} - \sum_{w \in U^{\text{heat}}_p} \sum_{\substack{k \in I^{\text{avail}}_{w,i} : \\ k \geq j+1}} X^{\text{dev}}_{b,w,k} \leq 1 \qquad \forall b \in B : \text{th} \notin G^{\text{imp}}_b, p \in P, \; u, v \in U^{\text{heat}}_p : z^{\text{dev}}_{u,v} = 1, i \in I^{\text{cov}}_{b,p}, \; h, j \in I^{\text{avail}}_{u,i} \setminus I^{\text{avail}}_{v,i} : h+1 \leq j \tag{4.88}$$

$$X^{\text{dev}}_{b,u,j} - \sum_{v \in U^{\text{heat}}_p} \sum_{k \in I^{\text{avail}}_{v,i}} X^{\text{dev}}_{b,v,k} \leq 0 \qquad \forall b \in B : \text{th} \notin G^{\text{imp}}_b, p \in P, u \in U^{\text{heat}}_p, \; i \in I^{\text{cov,inst}}_{b,p,u}, j \in I : j \leq i, j \notin I^{\text{avail}}_{u,i} \tag{4.89}$$

to be used within the Benders RMP. Constraints (4.87) ensure that at least one device $u \in U^{\text{heat}}_p$ needs to be installed in year $j \in I^{\text{avail}}_{u,i}$ for every year $i \in I^{\text{cov}}$ for buildings without district heating $b \in B : \text{th} \in G^{\text{imp}}_b$. However, a device $v \in U^{\text{heat}}_p$ with $l^{\text{dev}}_v < l^{\text{dev}}_u$ with $u \in U^{\text{heat}}_p$ where the installation of u satisfies constraints (4.87) for building $b \in B$ and year $i \in I$ for $p \in P$ might not be available in

i. For this reason, constraints (4.88) enforce the installation of at least one more device $w \in U_p^{\text{heat}}$ in the described situation such that the peak demand $p_{b,p}^{\text{peak}}$ can be covered. Similarly, constraints (4.89) enforce an installation of a device $v \in U_p^{\text{heat}}$ for $p \in P$ in building $b \in B : \text{th} \notin G_b^{\text{imp}}$ if the installation of a device $u \in U_p^{\text{heat}}$ lead to the deinstallation of an existing device and v is not available anymore in year $i \in I^{\text{cov,inst}} b, p, u$ such that $p_{b,p}^{\text{peak}}$ cannot be covered.

Benders RMP

The Benders RMP ($\text{RMP}^{\text{BOESP}}$) is given by

$$\min \quad \phi \left(C^{\text{inv,dev,fix}} + C^{\text{inv,mea}} + C^{\text{om,fix}} \right) + (1 - \phi)\, \omega\, TGW^{\text{mea}} + \sum_{b \in B} \eta_b \tag{4.90}$$

$$\text{s.t.} \quad (3.5) - (3.11) \text{ and } (3.13) - (3.15) \tag{4.91}$$

$$(3.11), (3.16) - (3.19), (3.58), (3.60) \text{ and } (3.62) - (3.64) \tag{4.92}$$

$$(\text{RMP}^{\text{BOESP}}) \quad (3.58), (3.60) \text{ and } (3.62) - (3.64),$$

$$(3.70), (3.74), (3.76), (3.78) - (3.80) \text{ and } (3.85) \tag{4.93}$$

$$(4.87) - (4.89) \tag{4.94}$$

$$\text{Benders cuts} \tag{4.95}$$

$$\eta_b \in \mathbb{R} \qquad \forall b \in B. \tag{4.96}$$

Continuous variables η_b in ($\text{RMP}^{\text{BOESP}}$) are used for the construction of Benders cuts in (4.95). Constraints in (4.91) – (4.93) pertain to constraints and variables for the allocation of resources at building stock level, building efficiency pathways and the calculation of costs and emissions, respectively. Constraints (4.94) are previously introduced valid inequalities that we use to avoid infeasibilities in problems ($\text{SP}_k^{\text{BOESP}}$). Benders cuts in (4.95) therefore only contain optimality cuts.

4.3.4 Building Operations, Energy System and Efficiency Pathways (BOESEP)

Benders SPs

In decomposition BOESEP, constraints and variable declarations corresponding to building efficiency pathways, building energy system pathways, and building operations are included in Benders SPs. Due to linked constraints in both building efficiency pathways and building energy system pathways, Benders SPs are only decomposable with regards to buildings $b \in B$. Therefore, like in decompositions BOEP and BOESP, we use also the partition of B into non-empty disjoint sets

$B_k \subseteq B$ for $k \in \{0, \ldots, n^{\text{sp}} - 1\}$ in this decomposition. We then define Benders SPs ($\text{SP}_k^{\text{BOESEP}}$) with

$$
(\text{SP}_k^{\text{BOESEP}}) \quad
\begin{array}{lll}
\min & \sum\limits_{b \in B_k^{\text{sp}}} O_b^{\text{sp}} & \quad (4.97) \\
\text{s.t.} & O_b^{\text{sp}} = \sum\limits_{i \in I} \Big(\phi \left(C_{b,i}^{\text{inv,dev,var}} + C_{b,i}^{\text{om,dev,var}} + C_{b,i}^{\text{en,sp}} \right) & \\
& \qquad + (1 - \phi)\, \omega\, TGW_{b,i}^{\text{op}} \Big) \qquad \forall b \in B_k^{\text{sp}} & \quad (4.98) \\
& (3.13) \text{ and } (3.14) \qquad \forall b \in B_k^{\text{sp}} & \quad (4.99) \\
& (3.16) - (3.19) \; \underline{\text{or } (3.19) \text{ and } (4.1) - (4.7)} \qquad \forall b \in B_k^{\text{sp}} & \quad (4.100) \\
& (3.21) - (3.32) \; \underline{\text{or } (3.23), (3.26) \text{ and } (3.27),} & \\
& \underline{(3.30) - (3.32) \text{ and } (4.8) - (4.23)} \qquad \forall b \in B_k^{\text{sp}} & \quad (4.101) \\
& (3.35) - (3.54) \; \underline{\text{or } (3.35) - (3.40), (3.43) - (3.45),} & \\
& \underline{(3.48) - (3.54) \text{ and } (4.24) - (4.31)} \qquad \forall b \in B_k^{\text{sp}} & \quad (4.102) \\
& (3.59), (3.65) \text{ and } (3.67), & \\
& (3.69), (3.75), (3.81), (3.83) \text{ and } (3.84) \qquad \forall b \in B_k^{\text{sp}} & \quad (4.103) \\
& O_b^{\text{sp}} \in \mathbb{R} \qquad \forall b \in B_k^{\text{sp}}, & \quad (4.104)
\end{array}
$$

where we assume linking variables $X_{b,u,i}^{\text{dev}}$ and $X_{b,m,i}^{\text{mea}}$ to be fixed to values of a solution of the Benders RMP of this decomposition and constraints in (4.99) – (4.103) to be set up for $b \in B : b \in B_k^{\text{sp}}$. Continuous variables O_b^{sp} defined in (4.104) measure the building-wise objective function in terms of (4.98). Constraints in (4.99) – (4.103) pertain to the declaration of linking variables as well as the constraints of building efficiency pathways, building energy system pathways, building operations, and calculations of costs and emissions, respectively. Here, underlined constraints denote optional modified constraints as presented in Sect. 4.2. The objective function of ($\text{SP}_k^{\text{BOESEP}}$) is the same as in ($\text{RMP}^{\text{BOESP}}$) since variables $Y_{b,e,i}^{\text{eff}}$ pertaining to building efficiency pathways do not have any objective coefficients unequal to zero in the original MILP model.

Benders RMP

The Benders RMP ($\text{RMP}^{\text{BOESEP}}$) is given by

$$(\text{RMP}^{\text{BOESEP}}) \quad \begin{aligned} \min \quad & \phi \left(C^{\text{inv,dev,fix}} + C^{\text{inv,mea}} + C^{\text{om,fix}}\right) + (1-\phi)\,\omega\, TGW^{\text{mea}} \\ & + \sum_{b \in B} \eta_b && (4.105) \\ \text{s.t.} \quad & (3.5) - (3.11) \text{ and } (3.13) - (3.15) && (4.106) \\ & (3.58), (3.60) \text{ and } (3.62) - (3.64) \\ & (3.70), (3.74), (3.76), (3.78) - (3.80) \text{ and } (3.85) \\ & && (4.107) \\ & (4.70) \text{ and } (4.87) - (4.89) && (4.108) \\ & \text{Benders cuts} && (4.109) \\ & \eta_b \in \mathbb{R} \qquad \forall b \in B. && (4.110) \end{aligned}$$

Continuous variables η_b are used for the generation of Benders cuts in (4.109). For the aforementioned reasons, the objective function (4.105) in ($\text{RMP}^{\text{BOESEP}}$) is equal to the objective function (4.90) in ($\text{RMP}^{\text{BOESP}}$). Constraints in (4.106) and (4.107) pertain to constraints of resource allocation at the building stock level and calculation of costs and emissions, respectively. Constraints in (4.108) are valid inequalities introduced in decompositions BOEP and BOESP and guarantee that Benders SPs ($\text{SP}_k^{\text{BOESEP}}$) are always feasible because mentioned infeasibilities are independent and respectively eliminated by both valid inequalities. Therefore, Benders cuts in (4.109) only contain optimality cuts.

Solution Approach and Implementation 5

To determine optimal transformation pathways for a residential building sector, we implement a solution approach that facilitates solving the model decompositions developed in Chap. 4. In particular, we implement a B&BC algorithm by separating Benders cuts in callbacks of the B&C algorithm of the commercial MILP solver Gurobi (Gurobi Optimization, LLC, 2023). With the aim to improve overall computation time, we implement different acceleration techniques based on the concepts introduced in Sect. 2.3, and further implement a construction heuristic for the Benders RMP.

All approaches developed in this chapter are implemented in Python 3.10. For the implementation and optimization of our MILP models throughout the B&BC process, we use the Gurobi 10.0 Python API and solver callbacks of Gurobi. Access to the Git Repository with the implemented solution approach can be granted upon request to the authors.

5.1 Outline of the Solution Approach

The implementation of our solution approach is depicted in the flow chart in Fig. 5.1. In our implementation, all computations concerning the Benders RMP are handled in the *main process* while Benders SPs are handled in *parallel subprocesses*. This approach enables repeatedly reoptimizing Benders SPs in a parallel fashion. We categorize methods in both the main process and parallel subprocesses into the two parts *Initialization* and *Optimization Process* which we describe in detail in Sect. 5.2 and 5.3, respectively. In the following, we give a brief outline of the solution approach and the interactions between the main process and parallel subprocesses.

R. Delorme, *Optimization of Transformation Pathways for the Residential Building Sector*, BestMasters, https://doi.org/10.1007/978-3-658-47466-9_5

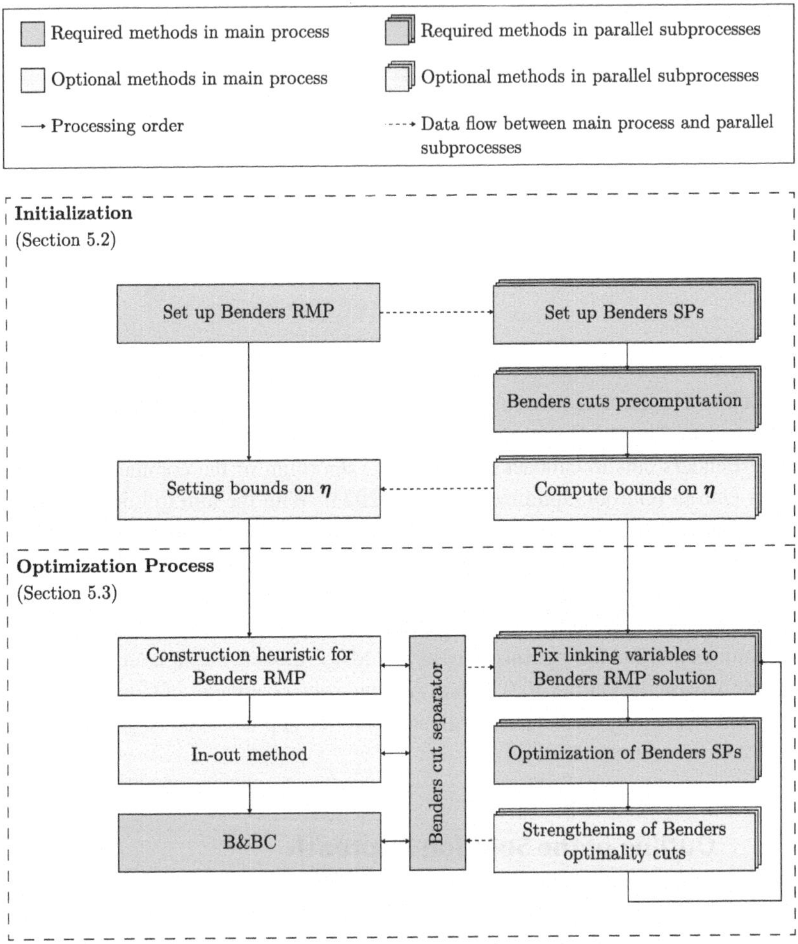

Fig. 5.1 Overview on the implemented method that facilitates the optimization of model decompositions within a B&BC algorithm. We categorize methods in the main process and parallel subprocesses into the two parts *Initialization* and *Optimization Process*

For the *Initialization*, the user first specifies the decomposition to be used for optimization. Depending on the chosen decomposition, the corresponding Benders RMP is set up. Subsequently, we create a parallel subprocess for each Benders SP. On each subprocess, we then precompute terms used for the generation of Benders cuts by identifying relevant constraints of the respective Benders SP in the corresponding coefficient matrix. Optionally, Benders SPs are then optimized without fixing linking variables to a Benders RMP solution to compute valid lower bounds on variables η in the Benders RMP. The precomputed terms are used in the main process to set up linear expressions with linking variables of the Benders RMP. In case bounds for variables η have been computed, those bounds are then established in the Benders RMP model in the main process.

The *Optimization Process* comprises methods that aim to determine an initial primal bound and improve the dual bound of the Benders RMP as well as the B&BC process. To determine an initial primal bound of the Benders RMP, we implement an optional construction heuristic by introducing a MILP model that determines primal bounds through solution cost estimates of the original model. Secondly, an optional in-out method can be used to improve the initial dual bound of the Benders RMP. During the in-out method, an LP relaxation of the Benders RMP is repeatedly used to separate Benders cuts. When the dual bound does not improve anymore, only tight Benders cuts are statically kept in the Benders RMP for the following B&BC process. The in-out method, therefore, aims to reduce the number of Benders cuts needed to be kept in the LP relaxation when leaving the root node and to shorten the B&BC tree. We regard both the construction heuristic and in-out method as optional methods since they are theoretically not needed to determine an exact solution to the problem. Lastly, in B&BC, we use callbacks of the MILP solver to retrieve the Benders RMP solution at a current node and separate Benders cuts. While we always separate Benders cuts for integer solutions of the Benders RMP, we leave it to the user if Benders cuts are separated for fractional solutions for nodes at different depths of the tree. The construction heuristic, the in-out method, and B&BC all require the separation of Benders cuts using the parallelized optimization of Benders SPs. Whenever a solution is available and Benders cuts need to be separated, a Benders cut separator is called, cf. Fig. 5.1, which passes the solutions to the respective subprocesses. When a solution is available in parallel subprocesses, linking variables of Benders SPs are fixed to the Benders RMP solution in the corresponding subprocess. Subsequently, Benders SPs are optimized by the simplex method, and the dual solution or the dual ray is retrieved. For Benders optimality cuts of decompositions BO and BOESP, Benders optimality cuts can optionally be strengthened by the procedure described in Sect. 4.3. Subsequently, the data used for Benders cut generation is returned to the main process where Benders cuts are generated using the precomputed linear terms. The subprocesses then idle until a new solution is available, ensuring that the simplex method can be warm started by using previously optimal solutions of the Benders SPs. The B&BC process terminates once the relative MILP gap falls below the predetermined value, cf. Sect. 2.2.1.

5.2 Initialization

In this section, we first explore the setup process of the Benders RMP and Benders SPs in parallel subprocesses. Subsequently, we briefly discuss the precomputation of linear expressions used for Benders cut generation in the main process. Lastly, we describe the process of computing bounds for variables η by optimizing Benders SPs without fixing linking variables to a Benders RMP solution.

5.2.1 Set Up

For the setup of the MILP models, the user may select one of the decompositions BO, BOEP, BOESP, and BOESEP. Furthermore, we include the possibility of further decomposing Benders SPs of decomposition BO by single years.

Benders RMP

The initialization starts by setting up the Benders RMP in the main process with the Gurobi Python API. For the Benders RMP, the user may choose whether to use valid inequalities for decompositions BOEP, BOESP, and BOESEP. Like Maher (2021), we use variable names to map solutions of the Benders RMP to Benders SPs. In the main process, we then create as many subprocesses as we have Benders SPs.

In the following, we describe the setup with parallel subprocesses and their interaction with the main process. To efficiently separate Benders cuts, we implement a parallelized Benders cut separation procedure in which computations of each Benders SP are performed on a single subprocess. For all decompositions, we choose to decompose Benders SPs by buildings $b \in B$, leading to a number of $n^{\text{sp}} = n^{\text{buil}}$ subprocesses. Furthermore, for decomposition BO, we add the option to further decompose Benders SPs by years $i \in I$, resulting in a number of $n^{\text{sp}} = n^{\text{buil}} n^{\text{years}}$ subprocesses.

Parallel Subprocesses and Benders SPs

Figure 5.2 shows how parallel subprocesses $k \in \{0, \ldots, n^{\text{sp}} - 1\}$ interact with the main process. Parallel subprocesses are created using the built-in `multiprocessing` package of Python 3.10. The `multiprocessing` package supports spawning processes through an API and offers local concurrency by establishing subprocesses, effectively enabling the user to more efficiently use multiple processors on a given machine (Python Software Foundation, 2023). First, we create a subprocess with the `Process` object of `multiprocessing` for each Benders SP $k \in \{0, \ldots, n^{sp} - 1\}$. For the interaction between the main process and subprocesses, we use two instances of the `Queue` object that is part of the `multiprocessing` module, which we henceforth call *task queue* and *result queue*. With the method `put`, the main process loads data into the task queue, while each subprocess uses the method `get` of the task queue to unload the provided data and then processes the data accordingly. If a single subprocess is finished with its respective job, it uses `put` to load the resulting data into the result queue. While subprocesses are processing the provided data, the main process is idling in the `get` method. Whenever a subprocess uses `put` to load processed data into the result queue, the main process unloads the data with `get` and processes the data, e.g., for Benders cut generation. The main process then continues with other tasks as soon as the last processed data from the result queue has been unloaded and processed.

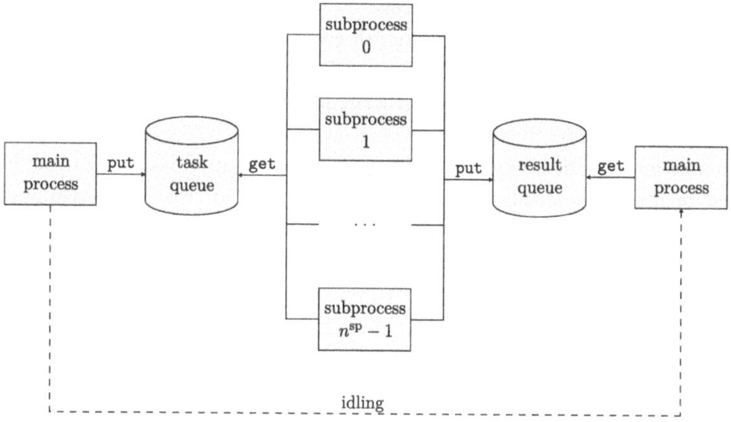

Fig. 5.2 Interaction between the main process and parallelized subprocesses using the task and result queues. Each Benders SP is handled by a single subprocess. Each subprocess idles until receiving data from the result queue, performs computations, and hands data back via the result queue

On each created subprocess, we then set up the Benders SPs according to the definitions of Benders SPs presented in Sect. 4.3. For all possible decompositions, we use the primal Benders SP presented in Sect. 4.3 for Benders cut generation. As explained in Sect. 2.1, if optimally solved by the simplex method, solving the primal version of an LP model also yields a dual solution. This dual solution can be directly retrieved after the model was solved to optimality by querying the respective constraints of the primal Benders SP. Similarly, in the case of infeasibility of a Benders SP, an extreme dual ray that proves the infeasibility of the Benders SP can be retrieved from the same constraints if the Parameter `InfUnbdInfo` is set to 1 (Gurobi Optimization, LLC, 2023).

5.2.2 Benders Cut Precomputation

To generate Benders cuts $\boldsymbol{\pi}_r(\mathbf{b} - \mathbf{B}\mathbf{y}) \leq 0$ in (2.32) or $\boldsymbol{\pi}_e(\mathbf{b} - \mathbf{B}\mathbf{y}) \leq \eta$ in (2.33), retrieving the dual solution and performing the algebraic calculations with the full dual vectors and matrices creates computational overhead since not the dual variables of all constraints are relevant to construct Benders cuts, cf. Sect. 4.2. Since in the whole optimization process we do not adjust the coefficient matrices of Benders SPs, it is possible to precompute linear expressions $b_j - \sum_{i=1} b_{j,i} y_i$ for a hopefully more efficient generation of Benders cuts.

Since the `gurobipy` module is not thread-safe, instances of objects `Var` in the `gurobipy` module cannot be passed between processes created with the `multiprocessing`. Therefore, Benders cuts must be constructed and added in the main process to the Benders RMP.

To precompute linear expressions $b_j - \sum_{i=1} b_{j,i} y_i$ used for Benders cut generation in the main process, we first identify the set of relevant constraints J for each Benders SP that are of interest for the construction of Benders cuts. For those relevant constraints J, we retrieve the corresponding values of the dual solution or dual ray after terminating the optimization process of the respective Benders SP. As explained in Sect. 4.2, the set of relevant constraints is $J = \{j \in \{1, \ldots, s\} : b_j \neq 0\} \cup \{j \in \{1, \ldots, s\} : \exists i \in \{1, \ldots, t\} : b_{j,i} \neq 0\}$, i.e., the set of constraints where linking variables have a matrix coefficient of unequal zero or the RHS is unequal to zero. For each relevant constraint $j \in J$, we then label a data structure with the unique constraint name and store the matrix coefficients $b_{j,i}$ of $\mathbf{B}$ with the corresponding variable names as well as the respective RHS b_j of $\mathbf{b}$. After identifying all relevant constraints and storing the corresponding matrix coefficients and RHSs, we return the data structure to the main process. Subsequently, in the main process, for each relevant constraint $j \in J$, we then construct linear expressions by using matrix coefficients with their respective variable names and retrieving the corresponding instances of the `Var` object of the Benders RMP. Whenever we separate a Benders cut by means of optimized Benders SPs, we retrieve the dual extreme point $\boldsymbol{\pi}_e$ or dual extreme ray $\boldsymbol{\pi}_r$ from Benders SPs and sum up precomputed linear expressions $b_j - \sum_{i=1} b_{j,i} y_i$ multiplied by the corresponding dual values $\pi_{e,j}$ or $\pi_{r,j}$ for Benders optimality and feasibility cuts, respectively.

5.2.3 Determining Bounds on η

By definition, variables η of the Benders RMP initially do not have a lower bound. Maher (2021) notes that for some problems, computing lower bounds for η before entering B&BC can be computationally beneficial. Bounds on η can be computed by optimizing Benders SPs without fixing linking variables to a Benders RMP solution. Note that we include trivial bounds for linking variables in Benders SPs as defined in the original model introduced in Chap. 3. Depending on the particular decomposition, we include additional constraints in Benders SPs to retain some of the restrictions imposed by constraints that are now included in the respective Benders RMP. Including additional constraints is important as otherwise retrieved objective values could pose useless lower bounds on variables η. For example, in the case of decomposition BO, (SP_k^{BO}) without fixed linking variables is unbounded from below since the available capacity for PV is not bounded in (SP_k^{BO}) and electrical energy can be sold to the grid without limitations. In Benders SP $k \in \{0 \ldots, n^{\text{sp}} - 1\}$ for decompositions BO and BOEP, we therefore add constraints

$$\dot{E}_{b,u,i}^{\text{avail}} \leq o_{b,u}^{\text{dev}} \qquad \forall b \in B_k^{\text{sp}}, u \in U, i \in I_k^{\text{sp}}. \tag{5.1}$$

Constraints (5.1) impose the upper limits $o_{b,u}^{\text{dev}}$ on the available device sizes for all years $i \in I$ and buildings $b \in B$. Thus, constraints (5.1) eliminate the case of unboundedness mentioned above. In Benders SP $k \in \{0 \ldots, n^{\text{sp}} - 1\}$ for decompositions BO and BOESP, we add constraints

$$\sum_{e \in E} Y_{b,e,i}^{\text{eff}} = 1 \qquad \forall b \in B_k^{\text{sp}}, i \in I_k^{\text{sp}}. \tag{5.2}$$

and therefore impose the existence of a thermal energy demand for all buildings and years. After optimization of the Benders SPs and retrieving the corresponding objective values, we return the objective values to the main process to bound η from below. Lastly, we remove constraints (5.1) and (5.2) from Benders SPs since we do not check those constraints when retrieving dual solutions or dual rays and leaving them thus could lead to the incorrect separation of Benders cuts if not considered in our Benders cut precomputation scheme.

5.3 Optimization Process

In this section, we first present the process of Benders cut separation by calling the Benders cut separator. Benders cuts are separated in three different methods that we subsequently cover in this section. First, we present an optional construction heuristic that determines an initial primal bound for the original problem. Second, we introduce the implementation of an optional in-out method that iteratively tightens an LP relaxation of the Benders RMP through Benders cut separation. Third, we briefly present the B&BC process in which Benders cuts are separated in solver callbacks of the B&C process of Gurobi.

5.3.1 Benders Cut Separation

To separate Benders cuts, a Benders RMP solution is handed to the Benders cut separator. From there, data is transferred to the Benders SPs in parallel subprocesses. Linking variables in Benders SPs are fixed, the Benders SPs are optimized and resulting Benders optimality cuts can then be strengthened, cf. Fig. 5.1, for decompositions BO and BOESP if applicable for the specific problem instance. Afterwards, the data used for Benders cut construction is transferred back to the main process where the Benders cut is constructed and added to the Benders RMP if violated.

5.3.2 Benders Cut Separator

To pass a Benders RMP solution to Benders SPs in parallel subprocesses, a data structure with the values of linking variables is loaded n^{sp} times into the task queue for each Benders SP. Afterwards, the main process awaits the results from parallel subprocesses, cf. Fig. 5.2. The Benders cut construction scheme works by the first in first out method. Therefore, whenever a dual solution or a dual ray of an optimized Benders SP is available in the result queue, the corresponding data structure is unloaded to the main process to construct a Benders cut using the cut precomputation scheme explained before. For the separation of Benders optimality cuts, we first check if the corresponding cut is violated. To account for numerical imprecision, we only construct and add the corresponding Benders optimality cut if $O_{b,i}^{\text{sp}*} \geq \bar{\eta}_{b,i} + \varepsilon^{\text{sep}}$ or $O_{b}^{\text{sp}*} \geq \bar{\eta}_{b} + \varepsilon^{\text{sep}}$ where $O_{b,i}^{\text{sp}*}$ and $O_{b}^{\text{sp}*}$ are the optimal objective values determined in Benders SPs for decomposition BO and decompositions BOEP, BOESP and BOESEP, respectively, and $\bar{\eta}_{b,i}$ and $\bar{\eta}_{b}$ are the corresponding values of underestimators in the current Benders RMP solution. For our implementation, we set $\varepsilon^{\text{sep}} = 1 \cdot 10^{-3}$.

We choose to await data of all parallel subprocesses and add all violated Benders cuts, even if a violated separated Benders cut has already been added to the Benders RMP. Therefore, the minimum time required for the separation and construction of Benders cuts is given by the highest processing time of all subprocesses. To reduce the overall computation time of our solution approach, we, therefore, aim to lower the highest processing time of all subprocesses as much as possible while keeping computational overhead coming with increased parallelization in acceptable ranges. We assume that the processing time of subprocesses mainly depends on the size of Benders SPs. Given the partitions of B and I into non-empty disjoint sets B_k^{sp} and I_k^{sp} for all $k \in \{0, \dots, n^{\text{sp}} - 1\}$, cf. Sect. 4.3, we set $|B_k^{\text{sp}}| = 1$ for all $k \in \{0, \dots, n^{\text{sp}} - 1\}$ for decompositions BOEP, BOESP and BOESEP. To evaluate the effect of disaggregated Benders SPs and cuts later, we set $|B_k^{\text{sp}}| = 1$ and $|I_k^{\text{sp}}| = 1$ for all $k \in \{0, \dots, n^{\text{sp}} - 1\}$ and $|B_k^{\text{sp}}| = 1$ and $|I_k^{\text{sp}}| = n^{\text{years}}$ for all $k \in \{0, \dots, n^{\text{sp}} - 1\}$ where we also aggregate Benders cuts by years $i \in I$ in a second implementation of decomposition BO.

Variable Fixing

Whenever a solution of the Benders RMP is available and Benders cuts need to be separated with Benders SPs, linking variables contained in the Benders SP need to be fixed to the values of linking variables of the current Benders RMP solution. We fix the values of linking variables to the Benders RMP solution by setting the lower and upper bound attributes `lb` and `ub` of the respective linking variable to the specified value. Like Maher (2021), we use shared variable names of linking variables in the Benders RMP and Benders SPs to map Benders RMP solutions to linking variables contained in Benders SPs.

Optimization

After fixing, the simplex method of Gurobi is used for the optimization of the respective Benders SP. By default, Gurobi solves the LP model by automatically choosing an algorithm. Typically, to solve LP models, Gurobi chooses a non-deterministic concurrent algorithm that concurrently employs the dual simplex method, barrier, and sometimes the primal simplex method, depending on the available number of available threads (Gurobi Optimization, LLC, 2023). By fixing linking variables in Benders SPs by means of adjusting the lower and upper bounds of the respective variable, the RHS of the coefficient matrix of the LP model is effectively changed. For a given solution of an LP model, the solution of the LP model stays dual feasible if the RHS of the primal LP model is changed since a change of the RHS corresponds to a change of the objective function of the dual LP model which thus does not modify the dual polyhedron. As changes in subsequent Benders RMP solutions in linking variables correspond to a change of the dual objective function, small changes in the Benders RMP solution also correspond to small changes in the dual objective function. Therefore, a new optimal dual basis might be found relatively close to the previously found dual optimum. Warm starting the dual simplex method, therefore, bears the potential to significantly reduce the time needed for reoptimization. Note that even though the LP model might be primal infeasible after fixing linking variables to a new Benders RMP solution, the dual basis can be converted to a primal basis with additional computational effort. Gurobi performs the warm start automatically if a model was previously solved to optimality and the `Model` object is not reset. Importantly, note that we do not employ a Benders cut selection technique that deals with the subject of potential dual degeneracy of Benders SP solutions. Since both methods might determine different optimal solutions, different Benders cuts might be found depending on whether Benders SPs are optimized by the primal or the dual simplex method. As a default, we select the dual simplex method instead of the concurrent algorithm to reduce computational overhead and still exploit the advantage of warm starting the reoptimization process directly.

Strengthening of Benders Optimality Cuts

When separating Benders optimality cuts for decompositions BO and BOESP, those Benders optimality cuts can be strengthened by the procedure introduced in Sect. 4.2 if mentioned conditions (C1) and (C2) hold for the respective problem instance. We only implement the Benders cut strengthening process for the case of modified constraints in Benders SPs as for Benders SPs with original constraints, the cut strengthening process is not as straightforward and does not fit well with the Benders cut precomputation scheme introduced earlier.

When using modified constraints, we assign new values to dual variables in each subprocess before passing the data structures to the main process. Each linking variable $Y_{b,e,i}^{\text{eff}}$ can be associated with the dual variable $\pi_{b,e,i}^{\text{eff}}$ that corresponds to the fixing constraint of that variable and, therefore, the strengthening process introduced in Sect. 4.2 can be applied as soon as the values of relevant dual variables have been retrieved. Instead of dual variable value $\overline{\pi_{b,e,i}^{\text{eff}}}$, we then store the value $\min_{\hat{e}\in W_{b,e,i}} \overline{\pi_{b,\hat{e},i}^{\text{eff}}}$ that corresponds to linear expression $-\mathrm{Y}_{b,e,i}^{\text{eff}}$ in the data structure to generate the strengthened Benders optimality cut (4.59). The new values are then used in the main process to generate strengthened Benders optimality cuts.

Conversely, when using the original constraints, linking variables $Y_{b,e,i}^{\text{eff}}$ appear in multiple linear expressions. For this reason, all Benders cut coefficients of single linking variables $Y^{\text{eff}}_{b,e,i}$ need to be aggregated before the cut strengthening procedure can be applied. Moreover, the dual solution cannot simply be manipulated like explained previously since Benders cuts need to be reconstructed in the main process given the introduced precomputation scheme. Note that Benders cuts could be strengthened in the main process which, however, leads to significantly more computational effort compared to the integration of the strengthening process in parallel subprocesses or by implementing a different secondary data structure for the reconstruction of Benders cuts in the main process.

5.3.3 Construction Heuristic

The first method that can be called in the optimization process is the construction heuristic to generate an initial primal bound, cf. Fig. 5.1. Note that, by design, Benders SPs are feasible for all solutions of the Benders RMP, cf. Sect. 4.3. Having an initial feasible solution of good quality might significantly reduce the size of the resulting B&BC tree and therefore reduce solving time. Note that the MILP solver itself employs different primal heuristics throughout the B&C process with the aim to improve the primal bound. However, depending on the size and complexity of the Benders RMP, finding a feasible solution can be challenging for general primal heuristics employed in the MILP solver. Therefore, using specific knowledge about the problem and tailoring a construction heuristic promises to be an effective way to efficiently determine an initial primal bound.

As a construction heuristic, we solve a new MILP model for which solutions can always be transformed into feasible solutions of the original model. In particular, we solve this MILP model by the regular B&C algorithm of Gurobi. In the model, we consider the case of $\phi = 1$ in the objective function (3.55) of the original model, i.e., the sole minimization of $-NPV$. Furthermore, we minimize discounted costs stemming from the coverage of thermal energy demand only, while including various assumptions for costs for investments, operations, and O&M. Precisely, we only consider the allocation of devices $u \in U^{\text{heat}}_{\text{th}}$ on buildings $b \in B$ in years $i \in I$. Since $U^{\text{heat}}_{\text{dhw}} \subset U^{\text{heat}}_{\text{th}}$, peak demands $p^{\text{peak}}_{b,p}$ for $p \in P$ can always be covered in all $b \in B : \text{th} \notin G^{\text{imp}}_b$ and years $i \in I$ if this is possible in the original model as well. Furthermore, if peak demands $p^{\text{peak}}_{b,p}$ for $p \in P$ are covered in all $b \in B : \text{th} \notin G^{\text{imp}}_b$ and $i \in I$ for a particular allocation of devices, one can always determine a feasible solution of the original model for that allocation since building energy systems of $b \in B : \text{th} \notin G^{\text{imp}}_b$ can always satisfy thermal energy demands at all time steps. Therefore, the MILP model of the construction heuristic is feasible if and only if the original model is feasible too. Moreover, we assume that costs in an optimal solution of the original model stem to a large degree from the costs connected to devices $u \in U_{\text{th}}^{\text{heat}}$. A reason for this is that for buildings $b \in B : \text{th} \notin G_b^{\text{imp}}$, devices $u \in U_{\text{th}}^{\text{heat}}$ need to be available in all years $i \in I$, entailing costs for investment, operation and O&M whereas electrical energy can directly be imported from the electricity grid. Apart from the allocation of devices $u \in U_{\text{th}}^{\text{heat}}$, we consider the realization of efficiency measures as in the original model. In the following, we introduce cost parameters for the new MILP model with which we estimate costs for solutions in the original model under various assumptions.

Cost Estimates

To find a primal bound for the original model of good quality, we try to estimate and approximate solution costs of the original model while significantly reducing model complexity. Those estimated costs are costs for the realization of building efficiency measures, costs for the installation of new devices, and costs for the operation of the building energy system.

For various cost estimations, we refer to the current building energy system design of building $b \in B$. For this, we define auxiliary cost parameters

$$c_{b,i,d,\tau}^{\text{exop,aux}} = \begin{cases} c_{\text{th},i}^{\text{en}} & \text{if th} \in G_{\text{th}}^{\text{imp}} \\ \sum\limits_{\substack{u \in U_{\text{th}}^{\text{heat}}:\\ l_{b,u}^{\text{exdev}} \geq 1}} \sum\limits_{g \in H_g^{\text{in}}} \dfrac{c_{g,i}^{\text{en}}}{\eta_{b,u,\text{th},d,\tau}} & \text{if } \exists\, u \in U : l_{b,u}^{\text{exdev}} \geq 1 \\ \dfrac{c_{\text{gas},i}^{\text{en}}}{\eta_{b,\text{BOI}^{\text{gas}},\text{th},d,\tau}} & \text{otherwise} \end{cases} \quad \forall b \in B, i \in I, d \in D, \tau \in T \tag{5.3}$$

that represent the undiscounted costs of one unit of thermal energy covered in year $i \in I$ on typical day $d \in D$ and hour $\tau \in T$ by the existing method to satisfy thermal energy demand in building $b \in B$. Note that by the problem definition, each building is connected to district heating, has a device installed that may cover thermal energy demand, or does not have a device installed because it has reached its lifetime. In case no device is installed, we define the auxiliary costs $c_{b,i,d,\tau}^{\text{exop,aux}}$ by the building energy system operation with device BOI^{gas} to approximate real costs.

The estimated costs for the realization of efficiency measures comprise investment costs for the realization of the efficiency measure and the difference in costs resulting from the operation of the energy system. To account for the change in building efficiency by means of the installation of efficiency measure $m \in M$, we define the singleton

$$E_{e,m} = \left\{\hat{e} \in E : z^{\text{eff}}_{m,\hat{e}} = 1, z_{\hat{m},\hat{e}} = 1 \ \forall c \in C^{\text{mea}} : m \notin M_c, \hat{m} \in M_c : z^{\text{eff}}_{\hat{m},e} = 1\right\} \quad \forall e \in E, m \in M \tag{5.4}$$

that contains the building efficiency $\hat{e} \in E$ that results from the realization of efficiency measure $m \in M$ for a given building efficiency $e \in E$. For the realized efficiency measure $m \in M$ we assume that it remains in the building for the rest of the optimization time horizon. The differences in building operation costs are estimated with respect to the existing building efficiency state e_b^{exeff} and building energy system design. The estimated discounted costs for the realization of an efficiency measure are then given by

$$\begin{aligned} c^{\text{mea,est}}_{b,m,i} = \max \Bigg\{ & \sum_{\substack{j \in I: \\ j \geq i}} \frac{1}{(1+\rho)^{j+1}} \sum_{e \in E_{e_b^{\text{exeff}},m}} \sum_{d \in D} \sum_{\tau \in T} w^{\text{day}}_{b,d} c^{\text{exop,aux}}_{b,i,d,\tau} \Delta^{\text{ts}} \left(p^{\text{th}}_{b,e,d,\tau} - p^{\text{th}}_{b,e_b^{\text{exeff}},d,\tau}\right) \\ & + c^{\text{inv,mea}}_{b,m,i} \frac{1 - h^{\text{mea}}_{u,i}}{(1+\rho)^i} c^{\text{inv,mea}}_{b,m,i}, \varepsilon^{\text{mea}} \Bigg\} \qquad \forall b \in B, m \in M, i \in I. \end{aligned} \tag{5.5}$$

Note that in (5.5) we introduce parameter $\varepsilon^{\text{mea}} > 0$ to guarantee that costs $c^{\text{mea,est}}_{b,m,i}$ are positive. Negative $c^{\text{mea,est}}_{b,m,i}$ would lead to the repeated realization of efficiency measures in solutions of our construction heuristic and thus an overestimation of savings in operational costs. On the other hand, we want to keep the model structure simple and therefore accept the deviation from true potential cost savings that could be realized with a single efficiency measure.

To estimate costs for the installation of a device $u \in U^{\text{heat}}_{\text{th}}$, we simply assume that the thermal energy output flow of those newly installed devices exactly covers the peak thermal demand of the respective building $b \in B$. The cost parameters are therefore given by

$$c^{\text{dev,est}}_{b,u,i} = \frac{1 - h^{\text{dev}}_{u,i}}{(1+\rho)^{i+1}} \left(c^{\text{dev,fix}}_{u,i} + p^{\text{peak}}_{b,\text{th}} c^{\text{dev,var}}_{u,i}\right) \quad \forall b \in B, u \in U^{\text{heat}}_{\text{th}}, i \in I. \tag{5.6}$$

Parameters for the estimation of building operation costs are given by

$$\begin{aligned} c^{\text{op,est}}_{b,u,i} = \frac{1}{(1+\rho)^{i+1}} \Bigg(& \sum_{g \in H^{\text{in}}_u} \sum_{d \in D} \sum_{\tau \in T} w^{\text{day}}_{b,d} c^{\text{en}}_{g,i} \Delta^{\text{ts}} \frac{p^{\text{th}}_{b,e_b^{\text{exeff}},d,\tau}}{\eta_{b,u,\text{th},d\tau}} \\ & + c^{\text{om}}_u \left(c^{\text{dev,fix}}_{u,i} + p^{\text{peak}}_{b,\text{th}} c^{\text{dev,var}}_{u,i}\right)\Bigg) \qquad \forall b \in B, u \in U^{\text{heat}}_{\text{th}}, i \in I \end{aligned} \tag{5.7}$$

where $c^{\text{op,est}}_{b,u,i}$ represent the discounted costs if device $u \in U^{\text{heat}}_{\text{th}}$ covers the entire thermal energy demand of building $b \in B$ and year $i \in I$ plus, and in deviation from the notation introduced for the model in Chap. 3, costs for O&M for device u. For the cost estimation in (5.7) we furthermore assume that the thermal energy demand of building b is defined by its initial building efficiency state e^{exeff}_b in all years $i \in I$.

Parameters to estimate costs for the operation with the current method to supply thermal energy demand are given by

$$c^{\text{curop,est}}_{b,i} = \frac{1}{(1+\rho)^{i+1}} \sum_{d \in D} \sum_{\tau \in T} w^{\text{day}}_{b,d} c^{\text{curop,aux}}_{b,i,d,\tau} \Delta^{\text{ts}} p^{\text{th}}_{b,e^{\text{exeff}}_b,d,\tau} \quad \forall b \in B, i \in I \tag{5.8}$$

where $c^{\text{curop,est}}_{b,i}$ are discounted costs to cover the thermal energy demand of building $b \in B$ by the currently existing method to satisfy heat demand in year $i \in I$.

Model

To formulate the MILP model to be used as the construction heuristic, we first define the parameters

$$z^{\text{cur}}_{b,i} = \begin{cases} 0 & \text{if th} \notin G^{\text{imp}}_b, i \notin I^{\text{cov}}_b \\ 1 & \text{otherwise} \end{cases} \quad \forall b \in B, i \in I \tag{5.9}$$

that are equal to one if and only if the current method to cover the thermal energy demand of building $b \in B$ can be used in $i \in I$. The MILP model is then given by

$$\min \sum_{b \in B} w^{\text{buil}}_b \sum_{i \in I} \Big(\sum_{m \in M} c^{\text{mea,est}}_{b,m,i} X^{\text{mea}}_{b,m,i} + \sum_{u \in U^{\text{heat}}_{\text{th}}} \big(c^{\text{dev,est}}_{b,u,i} X^{\text{dev}}_{b,u,i} + \sum_{j \in I^{\text{dev}}_{u,i}} c^{\text{op,est}}_{b,u,j} Y^{\text{avail}}_{b,u,i,j} \big) + c^{\text{curop,est}}_{b,i} Y^{\text{cur}}_{b,i} \Big) \tag{5.10}$$

$$\text{s.t.} \quad \sum_{u \in U^{\text{heat}}_{\text{th}}} \sum_{i \in I^{\text{avail}}_{u,j}} Y^{\text{avail}}_{b,u,i,j} + z^{\text{cur}}_{b,j} Y^{\text{cur}}_{b,j} = 1 \quad \forall b \in B, j \in I \tag{5.11}$$

$$X^{\text{dev}}_{b,u,i} - Y^{\text{avail}}_{b,u,i,j} \geq 0 \quad \forall b \in B, u \in U^{\text{heat}}_{\text{th}}, i \in I, j \in I^{\text{dev}}_{u,i} \tag{5.12}$$

$$\text{(CH)} \quad \sum_{v \in U^{\text{heat}}_{\text{th}}} \sum_{\substack{i \in I^{\text{avail}}_{u,j}: \\ i \neq j}} z^{\text{dev}}_{u,v} Y^{\text{avail}}_{b,v,i,j} + X^{\text{dev}}_{b,u,j} \leq 1 \quad \forall b \in B, u \in U^{\text{heat}}_{\text{th}}, j \in I \tag{5.13}$$

$$Y^{\text{avail}}_{b,u,i,j-1} - Y^{\text{avail}}_{b,u,i,j} \geq 0 \qquad \forall b \in B, u \in U^{\text{heat}}_{\text{th}}, i \in I, j \in I^{\text{dev}}_{u,i} \setminus \{i\} \tag{5.14}$$

$$(3.5) - (3.12) \tag{5.15}$$

$$(4.70) \text{ and } (4.87) - (4.89) \tag{5.16}$$

$$(3.13) - (3.15) \tag{5.17}$$

$$Y^{\text{avail}}_{b,u,i,j} \in [0, 1] \quad \forall b \in B, u \in U^{\text{heat}}_{\text{th}}, i \in I, j \in I^{\text{dev}}_{u,i} \tag{5.18}$$

$$Y^{\text{cur}}_{b,i} \in [0, 1] \quad \forall b \in B, i \in I \tag{5.19}$$

where $Y^{\text{avail}}_{b,u,i,j}$ and $Y^{\text{cur}}_{b,i}$ are non-negative continuous variables that take values between zero and one. The value of $Y^{\text{avail}}_{b,u,i,j}$ represents the share of thermal energy demand in year $j \in I$ and building $b \in B$ covered by device $u \in U^{\text{heat}}_{\text{th}}$ installed in year $i \in I^{\text{avail}}_{u,j}$. Similarly, the value of $Y^{\text{cur}}_{b,i}$ represents the share of thermal energy demand covered by the currently employed method to cover thermal energy demand in building $b \in B$ and year $i \in I$. Constraints in (5.17) are variable definitions for $X^{\text{dev}}_{b,u,i}$, $X^{\text{mea}}_{b,m,i}$ and $Y^{\text{encon}}_{b,i}$. In the objective function of (CH), we consider the investment costs into efficiency measures and devices with parameters $c^{\text{mea,est}}_{b,m,i}$ and $c^{\text{dev,est}}_{b,u,i}$. Costs for operations and O&M are considered with parameters $c^{\text{op,est}}_{b,u,i}$ and $c^{\text{curop,est}}_{b,i}$.

To estimate costs for building energy system operation, we use the modeling procedure similar to the one introduced in Sect. 3.3.3 to model energy system pathways. Constraints (5.11) ensure that the thermal energy demand is satisfied for all buildings $b \in B$ and years $i \in$ by using the current method for covering thermal energy demand or available energy supply from newly installed devices. Therefore, constraints (5.11) let us measure costs for building operations and O&M in the objective function. With constraints (5.12), we link the availability of device $u \in U^{\text{heat}}_{\text{th}}$ in years $j \in I^{\text{dev}}_{u,i}$ to the installation of the respective device in year $i \in I$. Next, for an installation of device $u \in U^{\text{heat}}_{\text{th}}$ in year $j \in I$, we use constraints (5.14) to disable the availability of devices installed in year $i \in I^{\text{avail}}_{j} : i \neq j$. Constraints (5.14) then ensure that those devices remain unavailable in following years. Lastly, in (5.15) and (5.16), we respectively use constraints pertaining to resource allocation at building stock level and valid inequalities introduced in Sect. 4.3 to ensure that a feasible solution can be transformed into a feasible solution for the original MILP model. Note that instead of sets U in constraints pertaining to the declaration of variables $X^{\text{dev}}_{b,u,i}$ (3.13) and constraints (3.5) and (3.7), we use sets $U^{\text{heat}}_{\text{th}}$ in sums and universal quantifiers.

Implementation

By construction, a feasible solution of (CH) can be transformed into a feasible solution of the original model and thus can be used to give a warm start to any Benders RMP. However, using the solution of (CH) with Gurobi is not straightforward when using callbacks during B&C. Variable objects can be assigned a `start` attribute with which Gurobi tries to compute an initial solution. However, if lazy constraints are added using callbacks during the computation of the initial solution, Gurobi 10.0.0 aborts the process and starts with regular B&C, regardless of whether separated lazy constraints are violated by the constructed solution (Gurobi Optimization, LLC, 2020). For this reason, we fall back to the procedure described in the following.

First, we relax the integrality requirements for all binary variables of the Benders RMP and then set the respective lower and upper bounds of those variables to the respective values in the solution of (CH) if the variable appears in (CH). If the relaxed binary variable does not appear in (CH), i.e., variables $X^{\text{dev}}_{b,u,i}$ $\forall b \in B, u \in U : u \notin U^{\text{heat}}_{\text{th}}, i \in I$, we set the respective lower and upper bounds to zero. Then, we solve the resulting LP model using separated Benders cuts to produce a full solution for the original Benders RMP. Note that the property of the encountered optimal solution for the Benders RMP with fixed variables is that separated Benders optimality cuts are not violated and thus not added to the Benders RMP. Subsequently, we store the solution in a data structure, restore the integrality requirement on variables and set their bounds to their original values. Then, we provide the solution from the data structure to all variables of the Benders RMP via the `start` attribute. Since the solution is feasible and no Benders optimality cuts are added to the model, Gurobi accepts the solution as an initial start solution and starts B&BC.

5.3.4 In-Out Method

Like the initial primal bound, a stronger initial dual bound may help reduce the size of the resulting B&BC tree and thus reduce overall computation time. Furthermore, when adding Benders cuts at the root node of the B&BC tree as lazy constraints, those Benders cuts are kept permanently in the LP relaxation of the Benders RMP even though potentially not being useful later on in the B&BC process. For those reasons, we employ an in-out method in which we relax the Benders RMP and iteratively separate Benders cuts that we add to the relaxation before entering B&BC. The implemented in-out method is shown in Algorithm 2 and is built upon the algorithmic implementations presented in Fischetti et al. (2017) and Maher (2021). In particular, like Fischetti et al. (2017), we employ the in-out method before entering the root node to tighten the LP relaxation of the Benders RMP. Conversely, we use the updating scheme of stabilization and separation points $\mathbf{y}^0$ and $\mathbf{y}^{\text{sep}}$ of Maher (2021) who employs the in-out method for Benders cut strengthening throughout the B&BC in the solver SCIP.

The implemented in-out method shown in Algorithm 2 uses a feasible primal solution as a stabilization point $\mathbf{y}^0$ if available and otherwise defines the stabilization point as a vector of zeros. Note that the primal solution can be constructed by our construction heuristic introduced before. After relaxing the integrality condition on integer variables of the Benders RMP, the algorithm enters a cut loop and the relaxed model is optimized to retrieve a first lower bound. With the resulting solution $(\bar{\mathbf{y}}, \bar{\eta})$, a separation point $\mathbf{y}^{\text{sep}}$ is assigned the convex combination $\lambda\bar{\mathbf{y}} + (1-\lambda)\mathbf{y}^0$. Subsequently, Benders cuts are separated for the point $\mathbf{y}^{\text{sep}}$ and added to the Benders RMP. If after k iterations the dual bound does not improve, we instead assign the separation point the vector $\lambda\bar{\mathbf{y}} + \boldsymbol{\delta}$ with $|\boldsymbol{\delta}| \ll 1$. After k more iterations without improvement of the dual bound, the separation point $\mathbf{y}^{\text{sep}}$ is assigned the Benders RMP solution $\bar{\mathbf{y}}$, commonly known as Kelley's cut separation scheme (Kelley, 1960). Lastly, after k more iterations without improvement of the dual bound, the cut loop terminates. Throughout the cut loop or when terminating the cut loop, we remove Benders cuts with slack every k iterations. The stabilization point $\mathbf{y}^0$ is updated after each separation of Benders cuts with $\mathbf{y}^0 \leftarrow \lambda\bar{\mathbf{y}} + (1-\lambda)\mathbf{y}^0$. Upon termination of the algorithm, we remove all Benders cuts with slack, statically add all tight Benders cuts to the Benders RMP, restore the integrality requirement on all integer variables of the Benders RMP, and enter B&BC.

5.3.5 B&BC

The last step of the optimization process is B&BC. While for integral Benders RMP solutions, a separation problem must be solved, the separation of Benders cuts is optional for fractional solutions of the Benders RMP, cf. Sect. 2.3.3. Like Maher (2021), we implement the possibility to generate Benders cuts for fractional Benders RMP solutions at different depths of the B&BC tree by introducing parameters d^{all} and d^{nth}. The parameter d^{all} indicates the node depth of the tree until which Benders cuts are always separated for fractional Benders RMP solutions. Furthermore, the parameter d^{nth} is used to separate Benders cuts for fractional Benders RMP solutions at node depths that are integer multiples of d^{nth}. Note that this implementation represents an adaptation of the two-phase method introduced by McDaniel and Devine (1977) when applied in the context of B&BC, cf. Sect. 2.3.3. When a Benders RMP solution at a node is integral or that node fulfills the node depth criteria specified by d^{all} and d^{nth}, the solution is passed to the Benders cut separator, cf. Fig. 5.1,

Algorithm 2 In-Out Method

```
if warm start solution $\mathbf{y}^{ws}$ is available then
    $\mathbf{y}^0 \leftarrow \mathbf{y}^{ws}$
else
    $\mathbf{y}^0 \leftarrow \mathbf{0}$
relax integrality condition on all integer variables of Benders RMP
$LB \leftarrow -\infty$
$n \leftarrow 0$
$l \leftarrow 0$
converged $\leftarrow$ False
while not converged do
    optimize Benders RMP and retrieve $z^{\text{RMP}}$ and $(\bar{\mathbf{y}}, \bar{\eta})$
    if $z^{\text{RMP}} \leq LB + \varepsilon^{\text{io}}$ then
        $n \leftarrow n + 1$
    else
        $n \leftarrow 0$
        $LB \leftarrow z^{\text{RMP}}$
    if $k \leq n < 2k$ then
        $\mathbf{y}^{\text{sep}} \leftarrow \lambda\bar{\mathbf{y}} + \boldsymbol{\delta}$
    else if $2k \leq n < 3k$ then
        $\mathbf{y}^{\text{sep}} \leftarrow \bar{\mathbf{y}}$
    else if $n = 3k$ then
        converged $\leftarrow$ True
    else
        $\mathbf{y}^{\text{sep}} \leftarrow \lambda\bar{\mathbf{y}} + (1 - \lambda)\mathbf{y}^0$
    separate Benders cuts for $(\mathbf{y}^{\text{sep}}, \bar{\eta})$ and add them as constraints to Benders RMP
    $\mathbf{y}^0 \leftarrow \lambda\bar{\mathbf{y}} + (1 - \lambda)\mathbf{y}^0$
    $l \leftarrow l + 1$
    if $l$ mod $k = 0$ or converged then
        if converged then
            optimize Benders RMP
        remove all Benders cuts in Benders RMP with slack greater than $\varepsilon^{\text{sl}}$
reintroduce integrality condition on all integer variables of Benders RMP
```

and Benders cuts are then separated in parallel subprocesses. While theoretically it is possible to add a non-empty subset of violated Benders cuts in a callback to discard the regarded Benders RMP solution, we decide to add all violated separated Benders cuts. We add all violated Benders cuts as lazy constraints to the Benders RMP model. In Gurobi, lazy constraints are directly added to the LP relaxation of the Benders RMP since they are necessary to describe the solution. However, Benders cuts separated for fractional solutions do not necessarily need to be added as lazy constraints but can be added as user cuts. Gurobi handles user cuts differently by applying filtering processes. While Gurobi therefore could discard some Benders cuts separated from fractional solutions that could help the optimization process later on, this procedure bears the potential to help keep the Benders RMP from growing excessively in size. For this reason, we experimentally evaluate the effect of adding Benders cuts for fractional Benders RMP solutions as user cuts in Chap. 6.

6 Computational Studies

To evaluate the performances of the different Benders decompositions introduced in Chap. 4 and the implementation with its acceleration techniques discussed in Chap. 5, we perform different computational studies. In Sect. 6.1, we introduce the framework with which we create instances of the problem and present the structure of our computational studies, the discussed performance measures, and used parameters. Subsequently, in Sect. 6.2, we present the results of our computational studies in which we investigate different configurational aspects of our model decompositions, the implementation, and the solver and evaluate the applicability of the solution approach to instances of larger sizes. Lastly, we summarize and discuss the results of our computational studies in Sect. 6.3.

6.1 Setup

In this section, we introduce different aspects of the setup of the computational study. First, we show how we generate instances of the introduced problem and define different instance sets used for computational studies. Subsequently, we present the structure of our computational studies and introduce metrics and performance measures that we use for the evaluation of computational results. Lastly, we give overviews on the different instance sets used in the computational studies, properties of the used hardware as well as parameters of the algorithm and solver.

Supplementary Information The online version contains supplementary material available at https://doi.org/10.1007/978-3-658-47466-9_6.

R. Delorme, *Optimization of Transformation Pathways for the Residential Building Sector*, BestMasters, https://doi.org/10.1007/978-3-658-47466-9_6

6.1.1 Instance Creation

To create instances for computational studies, values for all parameters and sets presented in Chap. 3 are required. As explained in Chap. 1, this thesis focuses on the development of a model and the algorithmic implementation of a Benders decomposition algorithm, leaving accurate parametrization for future research. However, to receive reliable performance benchmarks of the solution approach, we try to approximate such instances by making reasonable simplifications and estimates. For all instances, we use $\phi = 1$ in objective function (3.55). Furthermore, all instances satisfy conditions (C1) and (C2) needed for the strengthening of Benders optimality cuts. In the following, we give an overview of the procedure with which we generate different instances.

Building Properties

To construct instances, we use data on the German residential building stock to construct data describing archetype buildings. In particular, we use a data set of the Census 2011 in Germany in which residential buildings are statistically described on a regional level (Destatis, 2011). The household and building characteristics differentiated in the Census 2011 include properties like the number of apartments, the average apartment size, the year of construction, the building type, e.g., detached, semi-detached, or terraced, and the existing energy supply technology. To approximate regions R, we categorize the federal states of Germany into four regions. We choose the number of regions to be dependent on the number of buildings n^{buil} to be optimized in the building stock. Precisely, we determine the number of regions with $n^{\text{reg}} = \min\{4, \lceil \frac{n^{\text{buil}}}{20} \rceil\}$. Based on the number of regions included in the instance, we then determine a set of federal states constituting the set of used regions. The partition of federal states into regions is given in Table A.1 in the appendix in the electronic supplementary material. Based on the set of federal states, we then assign a federal state to each building $b \in B$ where we base the assignment probability on the total number of buildings of the respective federal states. Based on the data of the Census 2011, for each building $b \in B$, we then sample the year of construction, number of apartments, building size, and building type. Building geometries such as height, width, and roof area are set in accordance with the number of apartments, building size, and building type. The pitch and azimuth of the roof are randomly sampled from a uniform distribution.

Building Time Series

For each building $b \in B$, we consider climate data of test reference year (TRY) 2015 of DWD (2015). The data of TRY 2015 are representative climate time series data for one year and are provided with an hourly resolution and a spatial resolution of 1 km × 1 km. For each building $b \in B$, we use data from TRY 2015 pertaining to a grid cell of the respective administrative capital of the assigned federal state. The respective coordinates of those grid cells are presented in Table A.1 in the appendix in the electronic supplementary material. The data on direct solar irradiation in TRY 2015 is provided for a horizontal surface. Therefore, we convert the hourly direct horizontal irradiation into the hourly direct irradiation normal to the roof surface by considering the respective geographical position, i.e., latitude and longitude of the grid cell, as well the pitch and azimuth of the roof and the hour of the year.

For demand time series of electricity and thermal energy in the form of space heating and DHW, we use the free load profiles of the load profiles synPro of ISE (2022). The demand time series are given in quarter-hourly time steps and correspond to an average single-family house built between 1979 and 2001 with four inhabitants (ISE, 2022). We aggregate the time series to hourly time steps, scale the demand time series by the number of inhabitants living in the respective building $b \in B$ and apply hourly Gaussian noise to each time series.

The time series of temperature, converted direct solar irradiation, and the scaled demand time series are then clustered for a number of $n^{\text{days}} = 4$ typical days by a k-medoids clustering algorithm of Schütz et al. (2018) that the authors find to be well-suited for the clustering of time series in the context of building energy system optimization. Next to the clustered time series, the clustering algorithm of Schütz et al. (2018) also yields weights $w_{b,d}^{\text{days}}$. To derive time series $p_{b,e,d,\tau}^{\text{th}}$, we define a reference building efficiency $e^{\text{ref}} \in E$ for the now clustered space heating time series of ISE (2022). For each building efficiency $e \in E$, day $d \in D$ and time step $\tau \in T$, we scale the values $p_{b,e,d,\tau}^{\text{th}}$ of the clustered demand time series by the ratio of the building-dependent total thermal transmittance with respect to the reference building efficiency, ignoring intertemporal dynamics in thermal transmittance. The total thermal transmittance is specific to the building efficiency $e \in E$ in building $b \in B$ and is calculated by the respective building geometries, i.e., its heat transmitting areas, and thermal transmittances of efficiency measures $m \in M : z_{m,e}^{\text{eff}} = 1$ constituting building efficiency e. In particular, this procedure of determining parameters $p_{b,e,d,\tau}^{\text{th}}$ leads to instances satisfying condition (C2) presented in Sect. (4.3) that is needed for the application of the cut strengthening technique.

Initial Building Efficiency and Building Energy System Design

For each building $b \in B$, we use the data of the Census 2011 to determine the existing supply technology to satisfy thermal energy demand. However, as mentioned by Kotzur (2018), the Census 2011 differentiates by district heating, CHP supply, central heating, single-story heating, and apartment heating and is thus not directly applicable to our building energy system superstructure, cf. Sect. 3.1. Like Kotzur (2018), we use an additional data set of BDEW (2019) to further resolve the categories of central heating, single-story heating, and apartment heating. The data sets of BDEW (2019) contain statistical distributions for central heating and single-story heating with gas boilers, oil boilers, pellet boilers, and electrical heat pumps for all federal states of Germany in 2019. We then use data sets of the Census 2011 (Destatis, 2011) and BDEW (2019) to sample a heat supply technology for each building $b \in B$ depending on the respective federal state. In case no district heating is used in a building, i.e., $b \in B : \mathrm{th} \notin G_b^{\mathrm{imp}}$, the lifetime of the existing heat supply technology $l_{b,u}^{\mathrm{exdev}}$ is calculated by using the sampled year of building construction as the year of first installation for that device and assuming constant replacements of the device at the end of its lifetime l_u^{dev}.

The initial building efficiency e_b^{exeff} is sampled from all possible building efficiencies $e \in E$ from a uniform probability distribution. Furthermore, the remaining lifetime of efficiency measures $m \in M : z_{m,e_b^{\mathrm{exeff}}}^{\mathrm{eff}} = 1$ are sampled from a uniform probability distribution in $\{0, \ldots, l_m^{\mathrm{mea}} - 1\}$.

6.1.2 Structure of Computational Studies

In the following, we discuss the structure of our computational studies, for which we present the results in Sect. 6.2. Throughout this thesis, we introduced various possibilities to alter different configurations of the decomposition, the algorithm, and the MILP solver. Many of those configurations potentially interrelate with regards to the performance of the algorithm which is why their interplay should be considered when evaluating their impact on algorithm performance. However, a full factorial design to test all configurations is computationally intractable, especially for instances with high hardware requirements. For this reason, we group configurations of our solution approach into *model-based*, *algorithm-based*, and *solver-based* configurations. In the following, we use this grouping to perform computational studies in the three following stages, where in each stage, we evaluate the effect of different configuration realizations of the respective stage. We henceforth call such a realization of particular configurations a *solution strategy* (short *strategy*). The three stages are then used to incrementally refine strategies. In particular, the three stages pertain to

1. model-based configurations (Sect. 6.2.1)—In this first stage, we compare the performances of the decompositions BO, BOEP, BOESP, and BOESEP. In all decompositions, we choose to use modified constraints as well as valid inequalities, if applicable, as well as the warm start solution produced by the construction heuristic. We benchmark the performance of all decompositions against solving the original MILP model directly with Gurobi with and without the availability of the warm start solution. Lastly, we investigate the effect of modified constraints, valid inequalities, and the construction heuristic.
2. algorithm-based configurations (Sect. 6.2.2)—Here, we first evaluate the effect of adding cuts for fractional nodes of the B&BC tree that are deeper than the root node. Subsequently, we study the impact on performance when strengthening Benders optimality cuts. We finish this stage by evaluating the performance of applying the in-out method before entering B&BC.
3. solver-based configurations (Sect. 6.2.3)—In this last stage, we first evaluate the effect of adding Benders cuts separated at fractional nodes as user cuts instead of lazy constraints. Furthermore, we investigate the effect of choosing the primal simplex method to optimize Benders SPs.

Note that the grouping of configurations, the order of the stages as well as the particular parametrization of strategies within a stage are chosen heuristically. We first compare model-based configurations since we expect the choice of the decomposition to have a large impact on performances because of significant differences in the strengths of the LP relaxations of Benders RMPs and Benders cuts. Furthermore, we expect that potential performance improvements based on different configurations in the second or third stage impact strategies of previous stages in a similar manner.

Based on computational results in the third stage, we select a best-performing strategy to optimize large-scale instances in a last computational study in Sect. 6.2.4 and evaluate the transferability of previous results to larger instances. Furthermore, we compare the computational results against computational results attained when optimizing the aforementioned instances directly with the solver Gurobi.

6.1.3 Evaluation Measures and Metrics

To evaluate the results of our computational studies, we report different measures throughout this chapter. To compare the performance of different strategies on our test sets, we aggregate measures using the *shifted geometric mean*. Here, we use the definition of Achterberg (2008) who defines the shifted geometric mean $\text{sgm}_q(\boldsymbol{\alpha}_s)$ with shift q of performance measures $\boldsymbol{\alpha}_s = (\alpha_{1,s}, \ldots, \alpha_{n,s})^{\mathrm{T}}$ pertaining to strategy s by

$$\text{sgm}_q(\boldsymbol{\alpha}_s) = \left(\prod_{i=1}^{n}(\max\{1, \alpha_{i,s} + q\})\right)^{\frac{1}{n}} - q. \tag{6.1}$$

The shifted geometric mean reduces the impact of outliers compared to the arithmetic mean as well as the impact of very small values compared to the geometric mean and is the de-facto standard when comparing the average performance of different MILP solvers (Berthold and Hendel, 2021).

We specify the different symbols of evaluated measures, their description, and shift q, if applicable, in Table 6.1. Note that while in this chapter, we use aggregate metrics for performance evaluation, we provide instance-specific data in Appendix A.3 in the electronic supplementary material. For ease of notation, we henceforth denote $\overline{\boldsymbol{\alpha}_s}$ for the shifted geometric mean $\text{sgm}_q(\boldsymbol{\alpha}_s)$ with shift q being specified in Table 6.1.

Table 6.1 Symbols of evaluation measures for computational studies, their descriptions and shift for calculation of shifted geometric mean, if applicable

Measure	Description	q
$n^{\text{boc}}_{p,s}$	Number of Benders optimality cuts separated during optimization of problem instance p by strategy s	10
$n^{\text{bfc}}_{p,s}$	Number of Benders feasibility cuts separated during optimization of problem instance p by strategy s	10
$n^{\text{io,boc}}_{p,s}$	Number of Benders optimality cuts separated during the in-out method when optimizing problem instance p with strategy s	10
$n^{\text{io,add}}_{p,s}$	Number of tight Benders optimality cuts added to the Benders RMP after terminating the in-out when optimizing problem instance p with strategy s	10
$n^{\text{io,its}}_{p,s}$	Number of iterations in the in-out method when optimizing problem instance p with strategy s	10

(Continued)

Table 6.1 (Continued)

Measure	Description	q
$n_{p,s}^{\text{nodes}}$	Number of explored nodes during B&BC or standard B&C (Gurobi) when optimizing problem instance p with strategy s	10
$t_{p,s}^{\text{prim}}$	Time at which the best known primal bound is found during optimization of problem instance p with strategy s	10
$t_{p,s}^{\text{cb}}$	Total amount of time spent in solver callbacks during optimization of problem instance p with strategy s	10
$t_{p,s}^{\text{io}}$	Time spent in the in-out method when optimizing problem instance p with strategy s	10
$t_{p,s}^{\text{root}}$	Time spent at the root node during B&BC or standard B&C (Gurobi) when optimizing problem instance p with strategy s	10
$\overline{z}_p^{\text{best}}$ *	Best known primal bound for problem instance p	–
$\underline{z}_p^{\text{best}}$ *	Best known dual bound for problem instance p	–
$\overline{z}_{p,s}^{\text{lim}}$ **	Best known primal bound at time limit when optimizing problem instance p with strategy s	–
$\underline{z}_{p,s}^{\text{lim}}$ **	Best known dual bound at time limit when optimizing problem instance p with strategy s	–
$\Delta_{p,s}^{\text{lim}}$ ***	Relative MILP gap at the time limit when optimizing problem instance p with strategy s	1
$\Delta_{p,s}^{\text{root}}$ ***	Relative MILP gap when leaving the root node during B&BC or standard B&C (Gurobi) with respect to the best known primal bound for all strategies that optimize the given problem instance p	1

* We report those measures in Appendix A.2 in the electronic supplementary material.
** We report those measures in Appendix A.3 in the electronic supplementary material.
*** Conversion to percentages is applied after the calculation of the shifted geometric mean.

6.1.4 Instance Sets and Hardware

For our computational studies, we generate three instance sets. For comparability of different strategies, we choose the same hardware configuration for all strategies optimizing instances of the same instance sets. Basic properties of the instance sets as well as hardware configurations are shown in Table 6.2. For instances of the same instance set, we leave n^{buil} and n^{years} constant. Between instance sets, we vary the number of buildings n^{buil} to evaluate the transferability of results of computational studies to larger instance sizes. We adapt the amount of allocated memory to instance sizes contained in instance set, the most memory-demanding strategy to optimize those instances, and the number of threads used by Gurobi.

Table 6.2 Information on used instance sets for computational studies and hardware configurations for optimizations

Name	Instances	n^{buil}	n^{years}	CPUs	Memory (GB)
`10b_20y`	20	10	20	8	56
`20b_20y`	20	20	20	8	96
`50b_20y`	5	50	20	16	174

Simulations are performed with computing resources granted by RWTH Aachen University under project rwth1427. In particular, all computations were executed on the CLAIX-2018 high-performance cluster. The used CPUs are Intel Xeon Platinum 8160 Processors *SkyLake* with 2.1 GHz with hyperthreading disabled.

6.1.5 Gurobi Parameters

For our optimizations, we set some of the Gurobi parameters to values different from default values. Table 6.3 and 6.4 show an overview of the changed parameters for the original model, Benders RMPs, Benders SPs, and the model used for the construction heuristic presented in Sect. 5.3.3. To compare the B&BC algorithm against the performance of Gurobi, we set the same time limit and number of used threads for the standard model and the Benders RMP.

Table 6.3 Set Gurobi parameters for original models, Benders RMPs, Benders SPs and models used for the construction heuristic for instance sets `10b_20y` and `20b_20y`

Parameter	Orig. Model	Benders RMP	Benders SPs	Con. Heur.
`TimeLimit`	36000	36000	–	30
`LazyConstraints`	–	1	–	–
`InfUnbdInfo`	–	–	1	–
`Threads`	8	8	–	–
`Method`	–	–	1*	–

* We set this value to 0 for a computational study performed in Sect. 6.2.3

Table 6.4 Set Gurobi parameters for original models, Benders RMPs, Benders SPs and models used for the construction heuristic for instance set 50b_20y

Parameter	Orig. Model	Benders RMP	Benders SPs	Con. Heur.
TimeLimit	108000	108000	–	180
LazyConstraints	–	1	–	–
InfUnbdInfo	–	–	1	–
Threads	16	16	–	–
Method	–	–	1*	–

* We set this value to 0 for a computational study performed in Sect. 6.2.3

To be able to add lazy constraints callback and retrieve dual rays in case of infeasibility of primal Benders SPs, we set parameters LazyConstraints and InfUnbdInfo to 1 for the Benders RMP and the Benders SPs, respectively. Furthermore, by default, Gurobi sets the number of used threads to the number of available CPUs, while imposing a soft limit of 32 threads. Due to using batch jobs on nodes of the CLAIX-2018 high-performance cluster where the number of available CPUs exceeds 32, Gurobi would thus set the number of threads to 32. To avoid computational bottlenecks, e.g., the requirement of significantly more memory, we set the number of used threads to the number of CPUs specified in Table 6.3 and 6.4.

6.2 Results of Computational Studies

In this section, we present the results of our computational studies. To that end, we show the results of computational studies conducted in the three aforementioned stages that are based on changes considering model-based, algorithm-based and solver based configurations. Based on the results in the third stage, we select a best-performing strategy. We then present computational results for the optimization of large-scale instances with the previously best-performing strategy and compare those against results when optimizing with benchmark strategies.

6.2.1 Model-based Configurations

In this first stage, we show the results of our first computational studies. As explained in Sect. 6.1.2, the large amount of different possible strategies together with high hardware requirements make it necessary to heuristically reduce the number of explored strategies. As we expect the choice of the decomposition to have a large effect on the performance of the algorithm, we choose to study the impact of choosing different model decompositions for a fixed configuration first. Therefore, we define base strategies `bo_base`, `boep_base`, `boesp_base`, `boesep_base` for the decompositions BO, BOEP, BOESP and BOESEP, respectively. In the additional strategy `bo_dis_base`, we further disaggregate Benders SPs and Benders cuts by years $i \in I$, whereas in strategy `bo_base`, Benders SPs and Benders cuts are aggregated by buildings $b \in B$ only. The configurations of all base strategies, categorized by model-based, algorithm-based, and solver-based configurations, are presented in Table 6.5. We first evaluate the performance of the aforementioned base strategies and compare it against the performance of benchmark strategies `gurobi` and `gurobi_ws` that correspond to the optimization of the original MILP model with Gurobi without and with the application of a warm start solution constructed by our construction heuristic.

Table 6.5 Configuration of base strategies, categorized by model-based, algorithm-based, and solver-based configurations

Configuration	Value
Modified constraints	Yes
Valid inequalities (if applicable)	Yes
Compute Bounds on η	Yes
Warm start applied	Yes
Benders cuts for fractional solutions at root node	Yes
Benders cuts for fractional solutions for nodes deeper than root node	No
Strengthened Benders optimality cuts (if applicable)	No
In-out method applied	No
Add Benders cuts for fractional Benders RMP solutions as user cuts	No
Simplex method for optimization of Benders SPs	Dual

Comparison of the Base Strategies

In this first comparison, we use the seven strategies defined in the previous section to optimize instances in instance sets `10b_20y` and `20b_20y`. Table 6.6 and 6.7 show the shifted geometric means with shifts specified in Table 6.1 for measures $\mathbf{\Delta}_s^{\text{lim}}$, $\mathbf{\Delta}_s^{\text{root}}$, $\mathbf{t}_s^{\text{lim}}$, $\mathbf{t}_s^{\text{prim}}$ and $\mathbf{n}_s^{\text{nodes}}$.

Table 6.6 Average relative MILP gap at time limit, number of explored nodes, relative global MILP gap at the root node, time spent at the root node and time for reaching the best primal bound of base and benchmark strategies for the optimization of instances in 10b_20y

Strategy	$\overline{\mathbf{\Delta}_s^{\text{lim}}}$ (%)	$\overline{\mathbf{\Delta}_s^{\text{root}}}$ (%)	$\overline{\mathbf{t}_s^{\text{root}}}$ (s)	$\overline{\mathbf{t}_s^{\text{prim}}}$ (s)	$\overline{\mathbf{n}_s^{\text{nodes}}}$
bo_base	7.77	9.25	1033.88	8986.40	155331.25
bo_dis_base	4.48	7.61	1040.39	22354.17	66357.22
boep_base	15.10	13.24	459.24	437.07	136206.56
boesp_base	15.45	13.37	2562.86	10986.26	34411.23
boesep_base	18.18	15.15	1338.73	633.97	119226.41
gurobi	9.23	4.59	30794.21	26582.42	39.63
gurobi_ws	7.13	4.61	32736.62	16545.53	20.59

Table 6.7 Average relative MILP gap at time limit, number of explored nodes, relative global MILP gap at the root node, time spent at the root node and time for reaching the best primal bound of base and benchmark strategies for the optimization of instances in 20b_20y

Strategy	$\overline{\mathbf{\Delta}_s^{\text{lim}}}$ (%)	$\overline{\mathbf{\Delta}_s^{\text{root}}}$ (%)	$\overline{\mathbf{t}_s^{\text{root}}}$ (s)	$\overline{\mathbf{t}_s^{\text{prim}}}$ (s)	$\overline{\mathbf{n}_s^{\text{nodes}}}$
bo_base	9.62	9.60	2502.06	16168.82	81607.80
bo_dis_base	5.79	7.90	3257.17	25716.35	18449.50
boep_base	15.81	13.76	1147.11	370.91	54788.04
boesp_base	15.76	12.70	8858.97	1682.09	7184.57
boesep_base	17.59	14.16	2539.74	71.84	36535.66
gurobi	17.67	6.13	35198.56	20714.32	0.00
gurobi_ws	9.15	6.08	35578.79	4113.23	0.00

None of the strategies solve any of the instances of 10b_20y or 20b_20y to proven optimality. Our benchmark strategies gurobi and gurobi_ws spend most time at the root node, reaching an average relative root node gap $\overline{\mathbf{\Delta}_s^{\text{root}}}$ of approximately 4.6 %. This value is significantly smaller compared to strategies that apply Benders decomposition, where we find the largest root node gap $\overline{\mathbf{\Delta}_s^{\text{root}}}$ to be attained for boesep_base. The difference in relative root node gap $\overline{\mathbf{\Delta}_s^{\text{root}}}$ is not unexpected since strategies employing Benders decomposition, the Benders RMP does not contain all constraints of the original model and Benders cuts need to be separated at the root node to tighten the Benders RMP. Note that in decomposition BO, we only use constraints pertaining to building optimization in Benders SPs while Benders SPs of the other Benders decompositions contain more constraints of the original model. For this reason, Benders RMPs of decompositions BOEP, BOESP, and BOESEP relax the original model more than decomposition BO. The time the solver spends at the root node $\overline{\mathbf{t}_s^{\text{root}}}$ is significantly lower for base strategies employing Benders decomposition. A potential reason for the considerably lower values of $\overline{\mathbf{t}_s^{\text{root}}}$ of those strategies compared to strategies `gurobi` and `gurobi_ws` could be that the respective Benders RMPs are smaller in size than the original model and Gurobi thus spends less time at the root node with the aim to reduce the potential tree size. The time spent at the root node increases significantly for all strategies when optimizing instance set `20b_20y`.

For both instance sets `10b_20y` and `20b_20y`, the number of explored nodes $\overline{\mathbf{n}_s^{\text{nodes}}}$ is at least three orders of magnitudes lower for strategies `gurobi` and `gurobi_ws` than for base strategies employing Benders decomposition. In particular, for instances of `20b_20y`, the solver did not leave the root node in both strategies. Conversely, among the compared strategies, most nodes are explored by `bo_base` on both instance sets `10b_20y` and `20b_20y`. Remarkably, the number of explored nodes is significantly higher than for strategy `bo_dis_base`, probably due to the disaggregation of Benders cuts in `bo_dis_base`, leading to higher reoptimization times of the Benders RMP. Of all strategies compared here, strategy `bo_dis_base` yields the lowest relative MILP gaps $\overline{\mathbf{\Delta}_s^{\text{lim}}}$ at the time limit with about 4.5 % and 5.8 % on the respective instance sets `10b_20y` and `20b_20y`. Furthermore strategy `bo_base` yields slightly larger relative MILP gaps $\overline{\mathbf{\Delta}_s^{\text{lim}}}$ than `gurobi_ws` on both sets. Interestingly, the relative MILP gap $\overline{\mathbf{\Delta}_s^{\text{lim}}}$ increases significantly more for strategy `gurobi` between instance sets `10b_20y` and `20b_20y` compared to all other strategies. This might be a first indication that the performance of benchmark strategy `gurobi` does not scale well with instance size if a warm start solution is not used.

The quality of attained dual and primal bounds differs greatly between the different strategies. In Fig. 6.1, we show the relative difference of the dual bound to the best known dual bound $1 - \frac{\underline{z}_{p,s}^{\text{lim}}}{\underline{z}_p^{\text{best}}}$ for problem instance p over the relative difference of the primal bound to the best known primal bound $\frac{\overline{z}_{p,s}^{\text{lim}}}{\overline{z}_p^{\text{best}}} - 1$ for the same instance. Furthermore, Fig. 6.1 also shows those relative differences for the heuristic solution that is used for a warm start of all strategies except `gurobi`. Particularly, we also use the objective value corresponding to the found heuristic solution to calculate the relative difference in dual bounds. Note that the best dual and primal bounds $\underline{z}_{p,s}^{\text{best}}$ and $\overline{z}_{p,s}^{\text{best}}$ are determined considering all strategies s that optimize instance p. Of the strategies compared here, `bo_dis_base` consistently reaches strong dual bounds for instances of 10b_20y and for instances of 20b_20y, bo_dis_base determines all the best dual bounds of all strategies compared here. With respect to the relative difference in dual bounds, strategy bo_base determines weaker dual bounds than bo_dis_base for all instances. One reason for this discrepancy could be that disaggregated Benders optimality cuts applied in bo_dis_base are stronger than aggregated Benders optimality cuts of bo_base and thus lead to better dual bounds over time. Strategies related to other decompositions show on average significantly worse relative differences in dual bounds when compared with bo_base and bo_dis_base. A possible explanation for this is that compared to decomposition BO, decompositions BOEP, BOESP, and BOESEP relax the Benders RMP further and Benders cuts are not able to strengthen the LP relaxation of the Benders RMP as quickly. An indicator supporting this hypothesis is that the relative difference in dual bounds for boesep_base is even worse than for boep_base and boesp_base. Also note that compared to decompositions BO and BOEP, in decompositions BOESP and BOESEP, additional parts of the original objective function are projected out in Benders SPs. This may lead to the Benders RMP determining solutions that are further away from an optimal solution and those strategies are only able to retain this information slowly.

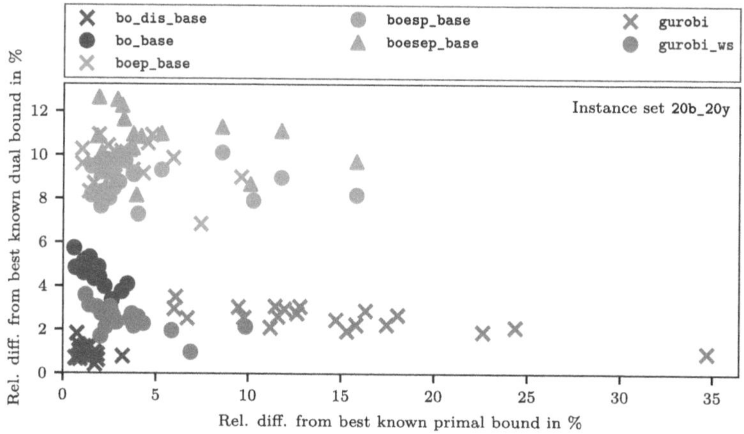

Fig. 6.1 Relative difference of dual and primal bounds with respect to the respective best known dual and primal bounds for strategies bo_base, bo_dis_base, boep_base, boesp_base, boesep_base, gurobi and gurobi_ws for instances of 10_b_20y and 20_b_20y. Furthermore, the relative differences of bounds for the warm start solutions are shown

When comparing all strategies except bo_dis_base, bo_base and gurobi with respect to the relative difference in primal bound, those differences are not as strongly pronounced as they are for the relative difference in dual bound. A conceivable explanation for this is that employed primal heuristics of Gurobi are similarly able to improve solutions of the Benders RMP for Benders decompositions or the original model for gurobi_ws. Note that irrespective of the Benders decomposition, for any encountered primal solution for the Benders RMP a feasible primal solution for the original model can be constructed if variables η are adapted to the bounds implied by separated Benders optimality cuts. Since the Benders RMP does not contain all constraints of the original model, primal heuristics that can improve or find new primal solutions for the original model should potentially also be able to determine comparable primal solutions for Benders RMPs. Strategies bo_dis_base and bo_base on average yield significantly better primal bounds than the other strategies. As we estimate primal heuristics to work comparably well for those decompositions, we believe a possible reason for the improvement is that stronger dual bounds throughout the B&BC process lead to better primal solutions being discovered within the branching process.

Furthermore, Fig. 6.1 gives information about the strength of the warm start solution of our construction heuristic. The highest relative difference of a warm start solution to the corresponding best-known dual bound is approximately 16 % for 10_b_20y and 20.5 % for 20_b_20y. For both instance sets, the majority of primal bounds given by warm start solutions are within a relative difference of 13 % from the best-known primal bound. While for instance set 10_b_20y strategy gurobi finds primal bounds of similar relative strength compared to our construction heuristic, primal bounds for instances of 20_b_20y are significantly weaker in average with respect to the relative difference to the best known primal bound.

Between strategies employing Benders decompositions, we find considerable differences with respect to the time spent in solver callbacks $\overline{\mathbf{t}_s^{\mathrm{cb}}}$ and the number of Benders optimality and feasibility cuts $\overline{\mathbf{n}_s^{\mathrm{boc}}}$ and $\overline{\mathbf{n}_s^{\mathrm{bfc}}}$. Table 6.8 and 6.9 show the shifted geometric means for time spent in solver callbacks $\overline{\mathbf{t}_s^{\mathrm{cb}}}$ and number of Benders cuts $\overline{\mathbf{n}_s^{\mathrm{boc}}}$ and $\overline{\mathbf{n}_s^{\mathrm{bfc}}}$ for base strategies employing Benders decomposition. First, note that no Benders feasibility cuts are separated due to the employment of valid inequalities In Benders RMPs of decompositions BOEP, BOESP, and BOESEP. The lowest average time in solver callbacks $\overline{\mathbf{t}_s^{\mathrm{cb}}}$ is achieved by bo_dis_base that only spends approximately 64 s in solver callbacks, a negligible fraction of the total solving time. For strategy bo_base, the time spent in solver callbacks is only slightly higher in absolute terms. A reason for strategy bo_dis_base for showing the lowest time in solver callbacks might be that this strategy disaggregates Benders SPs and contains the smallest LP sizes. Note that the minimum time spent in a single callback for Benders cut separation depends on the maximum time needed for the optimization of a Benders SP. For smaller Benders SPs, we expect the maximum time for reoptimization to be on average lower than for larger Benders SPs.

Table 6.8 Average time spent in solver callbacks and number of Benders optimality and feasibility cuts of base strategies for the optimization of instances in 10b_20y

Strategy	$\overline{\mathbf{t}_s^{\mathrm{cb}}}$ (s)	$\overline{\mathbf{n}_s^{\mathrm{boc}}}$	$\overline{\mathbf{n}_s^{\mathrm{bfc}}}$
bo_base	95.82	1496.62	0.00
bo_dis_base	64.54	14077.45	0.00
boep_base	283.71	2454.14	0.00
boesp_base	1194.30	3714.46	0.00
boesep_base	4978.58	5814.43	0.00

Table 6.9 Average time spent in solver callbacks and number of Benders optimality and feasibility cuts of base strategies for the optimization of instances in 20b_20y

Strategy	$\overline{\mathbf{t}_s^{\mathrm{cb}}}$ (s)	$\overline{\mathbf{n}_s^{\mathrm{boc}}}$	$\overline{\mathbf{n}_s^{\mathrm{bfc}}}$
bo_base	120.85	2792.63	0.00
bo_dis_base	230.31	26975.37	0.00
boep_base	209.58	3145.48	0.00
boesp_base	941.01	5267.52	0.00
boesep_base	2374.74	5174.97	0.00

Interestingly, both the time spent in solver callbacks and the number of callbacks differ significantly between instance sets 10b_20y and 20b_20y. While, strategies bo_base and bo_dis_base spend more time when optimizing instances of 20b_20y than 10b_20y, the opposite is the case for the other strategies. Possibly, finding primal solutions is harder for instances of 20b_20y and thus, the Benders cut separator is not called as often for all strategies. However, as hypothesized before, stronger dual bounds encountered in B&BC with strategies bo_base and bo_dis_base might lead to more primal solutions being found and thus the separator is called more often in relative terms. For all strategies except boesep_base, more Benders optimality cuts are separated on average for instances of 20b_20y, possibly due to the higher number of Benders SPs.

Modified Constraints and Valid Inequalities

To examine the effect of modified constraints and valid inequalities on the performance of the strategies, we define new strategies that differ from base strategies in the fact that they do not employ either modified constraints or valid inequalities if applicable for the decomposition. For the former case, we define strategies bo_nomod, bo_dis_nomod, boep_nomod and boesp_nomod and for the latter case strategies boep_novi and boesep_novi. Note that we do not introduce altered strategies for boesep_base as the underlying decomposition employs the same modified constraints or valid inequalities found in boep_base or boesp_base and we, therefore, assume that results are comparable to observations of changes of measures with respect to those base strategies.

Figure 6.2 shows values $\overline{\mathbf{\Delta}_s^{\text{lim}}}$, $\overline{\mathbf{t}_s^{\text{cb}}}$, $\overline{\mathbf{n}_s^{\text{bfc}}}$ and $\overline{\mathbf{n}_s^{\text{boc}}}$ for instances of 10b_20y for newly defined strategies in comparison with respective base strategies. None of the newly defined strategies solves any instance of 10b_20y to optimality. With relative respect to all aforementioned base strategies, not employing modified constraints increases time spent in solver callbacks, especially for strategies bo_nomod and bo_dis_nomod. Moreover, the number of separated Benders optimality cuts $\overline{\mathbf{n}_s^{\text{boc}}}$ decreases significantly when not employing modified constraints. Interestingly, for strategies bo_dis_nomod and boesp_nomod, the resulting MILP gap at time limit $\overline{\Delta_s^{\text{lim}}}$ decreases. In particular, bo_dis_nomod attains a relative MILP gap $\overline{\mathbf{\Delta}_s^{\text{lim}}}$ of 3.58 %, yielding the lowest value of $\overline{\mathbf{\Delta}_s^{\text{lim}}}$ until this point.

Leaving away valid inequalities leads to the separation of Benders feasibility cuts. However, the number of separated Benders feasibility cuts is approximately one order of magnitude lower than the number of separated Benders optimality cuts for strategies boep_novi and boesp_novi. Furthermore, for both strategies boep_nomod and boesp_nomod, the average relative MILP gap $\overline{\mathbf{\Delta}_s^{\text{lim}}}$ increases compared to boep_base and boesp_base.

In Fig. 6.3, we compare the relative difference of dual and primal bounds for strategies bo_dis_base and bo_dis_nomod with respect to the best known dual and primal bounds for instances of 10b_20y. Strategy bo_dis_nomod yields consistently lower dual bounds than bo_dis_base. We explain this with the lack of a cut selection mechanism for both strategies. As explained in Sect. 2.3.4, optimal solutions of Benders SPs are possibly dual degenerate, and therefore, more than one Benders cut exists that can cut off the Benders RMP solution. The modification of constraints by procedures (P1) and (P2) introduced in Sect. 4.2 changes Benders SPs and even though valid Benders optimality cuts may be separated, the corresponding dual solution may differ due to the way Benders SPs are optimized using the simplex method.

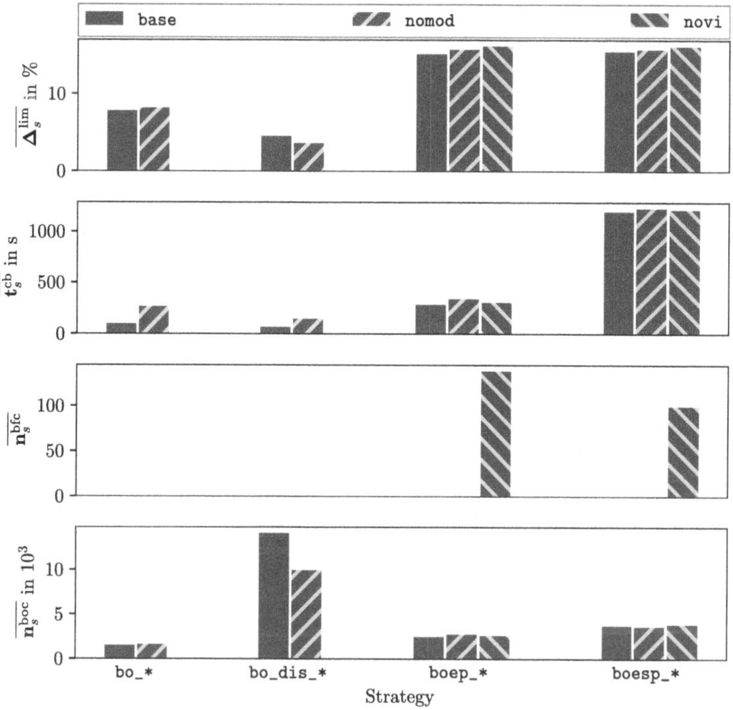

Fig. 6.2 Comparison of relative MILP gap at the time limit, time spent in solver callbacks and the number of separated Benders feasibility and optimality cuts for strategies `bo_nomod`, `bo_dis_nomod`, `boep_nomod`, `boesp_nomod`, `boep_novi` and `boesep_novi` and their corresponding base strategies

In the following section, we investigate changes to our strategies based on Benders decomposition by changing algorithm-based parameters. In particular, we will base newly defined strategies only on strategies `bo_dis_base`, `bo_dis_nomod` and `bo_base` and thus neglect other strategies because of their inability to compute dual bounds of similar strength.

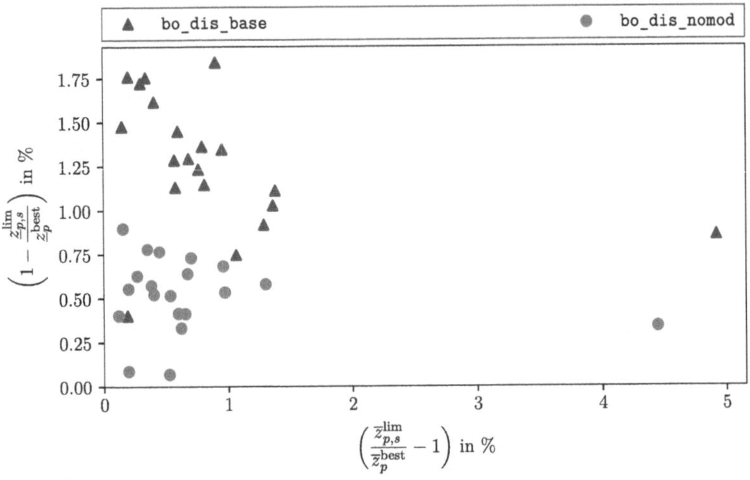

Fig. 6.3 Relative difference of dual and primal bounds with respect to the respective best known dual and primal bounds for strategies `bo_dis_base` and `bo_dis_nomod` for instances of `10_b_20y`

6.2.2 Algorithm-based Configurations

In this second stage, we conduct computational studies to evaluate the effect of changing algorithm-based configurations of strategies `bo_base`, `bo_dis_base` and `bo_dis_nomod`. First, we evaluate the effect of separating Benders cuts for fractional solutions for nodes deeper than the root node. In a second computational study, we then investigate the change in performance when strengthening Benders optimality cuts by the strengthening technique introduced in Sect. 4.3. Based on the best-performing strategies of those computational studies, we then evaluate the impact of tightening the Benders RMP using the in-out method introduced in Sect. 5.3.4 before entering the root node of B&BC.

Additional Separation of Benders Cuts

Separating Benders cuts for fractional solutions may help reduce the size of the B&BC tree and thus may lead to stronger dual bounds within the same time, but may conversely significantly increase the size of the Benders RMP and thus slow down the optimization. Here, we investigate the effect of separating Benders cuts for fractional solutions by setting $d^{\text{nth}} = 20$, i.e., by separating Benders cuts for fractional solutions at node depths that are integer multiples of 20. We define strategies bo_fr20, bo_dis_fr20 and bo_dis_nomod_fr20 that only differ in the aforementioned change from the respective strategies bo_base, bo_dis_base and bo_dis_nomod.

In Fig. 6.4, we compare values of $\overline{\mathbf{\Delta}_s^{\text{lim}}}$, $\overline{\mathbf{n}_s^{\text{nodes}}}$, $\overline{\mathbf{t}_s^{\text{cb}}}$, $\overline{\mathbf{t}_s^{\text{prim}}}$ and $\overline{\mathbf{n}_s^{\text{boc}}}$ for aforementioned strategies on instances of 10b_20y. None of the instances is solved to proven optimality by the newly defined strategies. Concerning strategies bo_base, bo_dis_base and bo_dis_nomod, separating Benders cuts with $d^{\text{nth}} = 20$ leads to a significantly larger number of separated Benders optimality cuts and time spent in solver callbacks, while the number of explored nodes is significantly

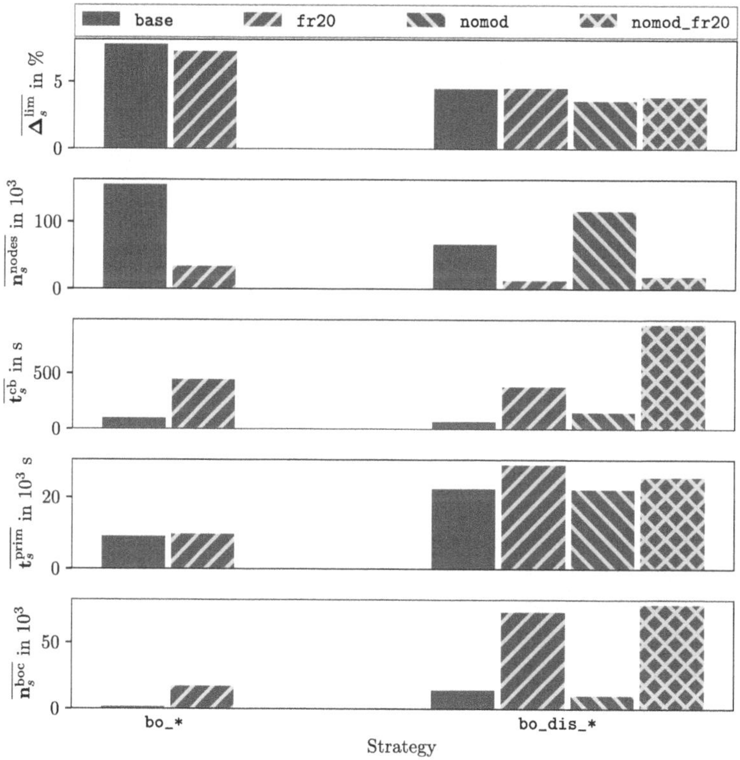

Fig. 6.4 Comparison of shifted geometric means of relative MILP gap at the time limit, number of explored nodes, time spent in solver callbacks, time to find the best primal bound and number of Benders optimality cuts for strategies bo_base, bo_fr20, bo_dis_base, bo_dis_fr20, bo_dis_nomod and bo_dis_nomod_fr20 for instances of 10b_20y

reduced. The only strategy that shows a decrease in $\overline{\mathbf{\Delta}_s^{\text{lim}}}$ with respect to its respective base strategy is bo_fr20. Notably, the time to find the best primal bound of the respective optimization $\overline{t_s^{\text{prim}}}$ increases when separating additional Benders optimality cuts for fractional solutions, especially for strategies based on decomposition BO with disaggregated Benders SPs and cuts. We hypothesize two reasons for this effect. First, separating significantly more Benders optimality cuts might slow down the optimization of Benders RMP LP relaxations and therefore retard the time a comparably good primal bound is found during B&BC. Second, additionally separated Benders cuts further down in the B&BC tree might tighten Benders RMP LP relaxations in such a way that more integer solutions are discovered by branching.

Significantly, since relative MILP gaps $\overline{\mathbf{\Delta}_s^{\text{lim}}}$ of strategies bo_base and bo_fr20 remain at significantly larger values than strategies bo_dis_base, bo_dis_fr20, bo_dis_nomod and bo_dis_nomod_fr20, we choose to not further refine strategies bo_base and bo_fr20 and focus on the remaining strategies instead.

Strengthening Benders Optimality Cuts

In Sect. 4.3, we present a technique to strengthen Benders optimality cuts that can be employed if conditions (C1) and (C2) are fulfilled by the respective problem instance. To evaluate the effect of the strengthening of Benders optimality cuts, we define new strategies bo_dis_str and bo_dis_fr20_str where the respective difference to strategies bo_dis_base and bo_dis_fr20 lies within the sole additional strengthening of Benders optimality cuts.

Figure 6.5 shows the relative difference of primal and dual gaps with respect to the respective best-known dual and primal bounds for strategies bo_dis_base, bo_dis_str, bo_dis_fr20 and bo_dis_fr20_str for instances of 10b_20y. None of the newly defined strategies solves any instance of 10b_20y to optimality. On average, the strengthening of Benders cuts improves the determined dual bound for the considered strategies. This is especially true for strategies bo_dis_base and bo_dis_str where dual gaps are improved with great consistency. The relative MILP gaps $\overline{\mathbf{\Delta}_s^{\text{lim}}}$ of strategies bo_dis_str and bo_dis_fr20_str are 4.04 % and 4.17 %, respectively, and thus improve the relative MILP gaps of strategies bo_dis and bo_dis_fr20. Note, however, that the relative MILP gap of bo_dis_str is still higher than for bo_dis_nomod on instances of 10b_20y. Since strengthened Benders cuts consistently improve the attained relative MILP gaps for a strategy, we only consider strategies bo_dis_str, bo_dis_fr20_str, bo_dis_nomod and bo_dis_nomod_fr20 for the application of the in-out method in the following.

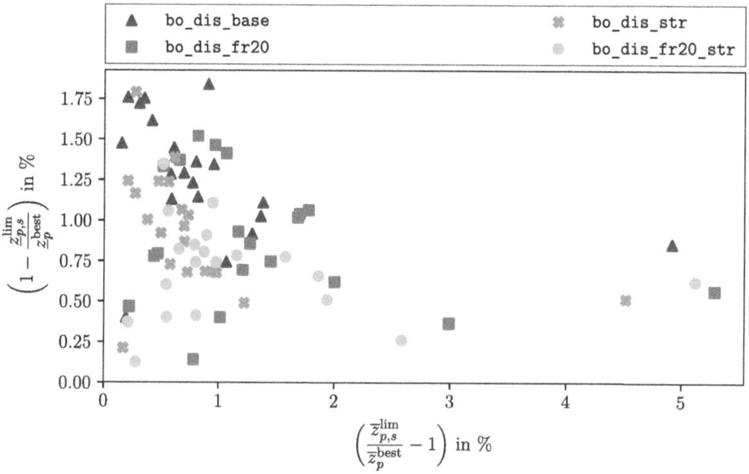

Fig. 6.5 Relative difference of primal and dual gaps of strategies bo_dis_base, bo_dis_str, bo_dis_fr20 and bo_dis_fr20_str for instances of 10b_20y

In-Out Method

In this section, we investigate the changes in performance measures when applying the in-out method to tighten the Benders RMP before entering the root node. For this, we define new strategies bo_dis_str_io, bo_dis_fr20_str_io, bo_dis_nomod_io and bo_dis_nomod_fr20_io that differ from strategies bo_dis_str, bo_dis_fr20_str, bo_dis_nomod and bo_dis_nomod_fr20 in the sole application of the in-out method before starting B&BC. We employ the in-out method shown in Algorithm 2, used with parameters $\lambda = 0{,}5$, $\boldsymbol{\delta} = (1 \cdot 10^{-6}, \ldots, 1 \cdot 10^{-6})^{\mathrm{T}}$ and $k = 5$. The time limit of 36 000 s for B&BC remains unchanged, however, we report the time spent in the in-out method.

Table 6.10 shows the time spent in the in-out method $\overline{\mathbf{t}_s^{\mathrm{io}}}$, the number of separated Benders optimality cuts in the in-out method $\overline{\mathbf{n}_s^{\mathrm{io,boc}}}$, the number of statically added Benders optimality cuts $\overline{\mathbf{n}_s^{\mathrm{io,add}}}$ and the number of in-out iterations $\overline{\mathbf{n}_s^{\mathrm{io,it}}}$ for the newly introduced strategies. Between strategies bo_dis_str_io and bo_dis_fr20_str_io as well as bo_dis_nomod_io and bo_dis_ fr20_io, respectively, values shown in Table 6.10 are not significantly different as those strategies do not differ from each other in methods before B&BC. However, strategies bo_dis_nomod_io and bo_dis_nomod_fr20_io require less iterations $\overline{\mathbf{n}_s^{\mathrm{io,it}}}$ and less Benders optimality cuts $\overline{\mathbf{n}_s^{\mathrm{boc,io}}}$ to converge than strategies bo_dis_str_io or bo_dis_fr20_str_io. Conversely, the time spent in the in-out method $\overline{\mathbf{t}_s^{\mathrm{io}}}$ is slightly greater for those strategies. This forms a consistent picture with the behavior previously observed in Sect. 6.2.1 where less separated Benders optimality cuts with a larger time spent in solver callbacks during B&BC lead to a stronger dual bound. We hypothesize that the additional time observed here might stem from the additional time to optimize Benders SPs, the additional time to construct Benders optimality cuts, and potentially higher reoptimization times for the Benders RMP using stronger Benders optimality cuts.

Table 6.10 Average time spent in the in-out method, number of separated Benders optimality cuts, number of statically added Benders optimality cuts at termination and number of in-out iterations for newly defined strategies employing the in-out method for the optimization of instances in 10b_20y

Strategy	$\overline{\mathbf{t}_s^{\mathrm{io}}}$ (s)	$\overline{\mathbf{n}_s^{\mathrm{io,boc}}}$	$\overline{\mathbf{n}_s^{\mathrm{io,add}}}$	$\overline{\mathbf{n}_s^{\mathrm{io,it}}}$
bo_dis_str_io	200.41	20441.77	200.00	103.22
bo_dis_fr20_str_io	211.06	20547.53	200.00	103.75
bo_dis_nomod_io	227.34	18995.82	200.00	95.98
bo_dis_nomod_fr20_io	238.75	18803.08	200.00	95.02

Note that for all strategies considered here, exactly 200 Benders optimality cuts are statically added in all instances to the Benders RMP when terminating the in-out method. This is exactly the minimum number of Benders cuts to be added to the Benders RMP when terminating the in-out method, since for instances of 10b_20y, we employ 200 Benders SPs and we know that no Benders SP is unbounded from below. Given the fact that the separated numbers of Benders optimality cuts $\overline{\mathbf{n}_s^{\mathrm{boc,io}}}$ is about two orders of magnitudes higher, the removal of Benders optimality cuts that are not tight in the last RMP optimum significantly reduces the size of the LP relaxation of the Benders RMP when entering the root node of the B&BC tree.

Figure 6.6 compares $\overline{\mathbf{\Delta}_s^{\mathrm{lim}}}$, $\overline{\mathbf{n}_s^{\mathrm{nodes}}}$, $\overline{\mathbf{t}_s^{\mathrm{cb}}}$ and $\overline{\mathbf{n}_s^{\mathrm{boc}}}$ of strategies bo_dis_str_io, bo_dis_fr20_str_io, bo_dis_nomod_io and bo_dis_nomod_fr20_io with the respective strategies that do not employ the in-out method.

The employment of the in-out method leads to lower relative MILP gaps $\overline{\boldsymbol{\Delta}_s^{\text{lim}}}$ in all four strategies. This decrease is slightly more pronounced for strategies `bo_dis_str_io` and `bo_dis_fr20_str_io` than for strategies `bo_dis_nomod_io` and `bo_dis_nomod_fr20_io`. Furthermore, the number of explored nodes $\overline{\mathbf{n}_s^{\text{nodes}}}$ is significantly increased in all strategies except `bo_dis_str_io`. This increase in explored nodes might be caused by less separated Ben-

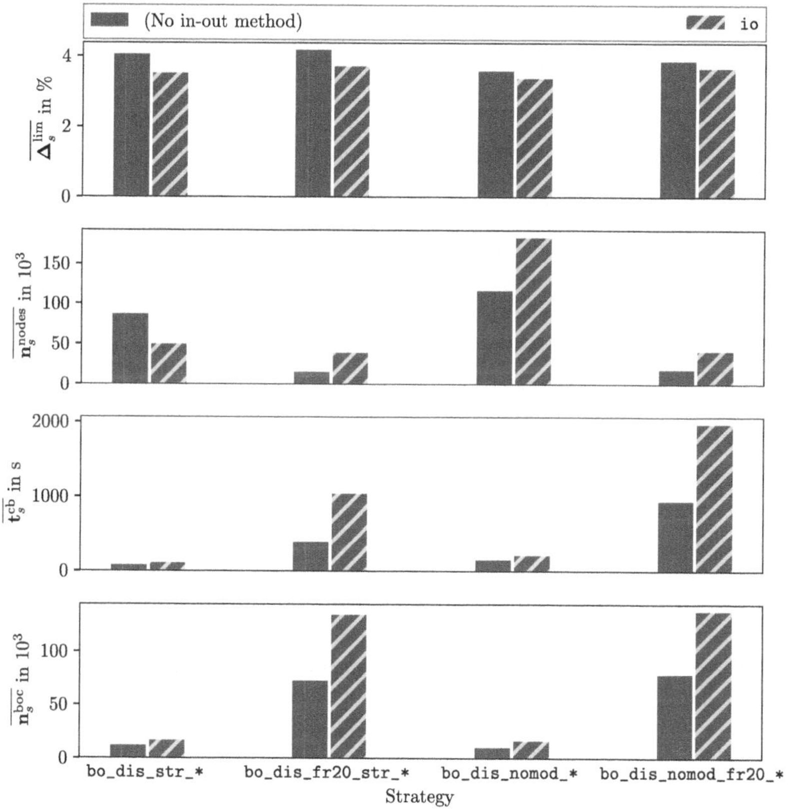

Fig. 6.6 Comparison of shifted geometric means of relative MILP gap at the time limit, number of explored nodes, time spent in solver callbacks and number of Benders optimality cuts in B&BC for strategies `bo_dis_str_io`, `bo_dis_fr20_str_io`, `bo_dis_nomod_io` and `bo_dis_nomod_fr20_io` as well as the respective strategies that do not employ the in-out method. The strategies optimize instances of `10b_20y`. None of the instances is solved to proven optimality

ders cuts being added to the Benders RMP until leaving the root node when applying the in-out method. In those cases, the Benders RMP relaxation might then be re-optimized faster, enabling the exploration of more nodes. Lastly, the employment of the in-out method increases the time spent in solver callbacks $\overline{\mathbf{t}_s^{\mathrm{cb}}}$ and the number of separated Benders cuts $\overline{\mathbf{n}_s^{\mathrm{boc}}}$ in B&BC significantly, especially for strategies `bo_dis_fr20_str_io` and `bo_dis_nomod_fr20_io`.

As the employment of the in-out method consistently decreases the relative MILP gap $\overline{\boldsymbol{\Delta}_s^{\mathrm{lim}}}$ for all considered strategies, we only consider strategies `bo_dis_str_io`, `bo_dis_fr20_str_io`, `bo_dis_nomod_io` and `bo_dis_nomod_fr20_io` for the following computational studies.

6.2.3 Solver-based Configurations

Here, we present the results of computational studies of the third stage in which we briefly evaluate the effect of two parameters of Gurobi on optimization performance. First, we evaluate the impact of adding Benders optimality cuts separated at fractional solutions as user cuts instead of lazy constraints. In a second computational study, we then optimize Benders SPs using the primal simplex method.

User Cuts and Lazy Constraints

For all strategies introduced until this point, we add all separated Benders optimality cuts separated at fractional solutions as lazy constraints during solver callbacks. However, since Benders cuts only need to be separated for integer solutions of the Benders RMP, adding Benders optimality cuts separated from fractional solutions as lazy constraints might significantly increase the size of the Benders RMP and thus slow down the optimization process. When adding constraints as user cuts, Gurobi applies internal filters to the respective user cuts and decides if and how to use them to improve the dual bound. While Gurobi thus might discard Benders optimality cuts that could be useful in improving the dual bound at a node of the B&BC tree, this approach could help keep the size of the Benders RMP smaller and thus inhibit an excessive increase of reoptimization time for the LP relaxation of the Benders RMP.

To evaluate the effect of adding Benders cuts separated at fractional solutions of the Benders RMP as user cuts instead of lazy constraints, we define strategies `bo_dis_str_io_uc`, `bo_dis_fr20_str_io_uc`, `bo_dis_nomod_io_uc` and `bo_dis_nomod_fr20_io_uc` that differ from strategies `bo_dis_str_io`, `bo_dis_fr20_str_io`, `bo_dis_nomod_io` and bo_dis_nomod_fr20_io in only that change of configuration. With the newly defined strategies, we then optimize instances of 10b_20y. None of the newly defined strategies solves any instance of 10b_20y to proven optimality.

In Fig. 6.7, we compare the relative MILP gap $\overline{\mathbf{\Delta}_s^{\text{lim}}}$ at the time limit, the number of explored nodes $\overline{\mathbf{n}_s^{\text{nodes}}}$ and the number of separated Benders optimality cuts $\overline{\mathbf{n}_s^{\text{boc}}}$ of strategies bo_dis_str_io_uc, bo_dis_fr20_str_io_uc, bo_dis_nomod_io_uc and bo_dis_nomod_fr20_io_uc with respective strategies that do not add Benders cuts separated at fractional Benders RMP solutions as user cuts. Adding separated Benders cuts separated at fractional solutions as user cuts instead of lazy constraints leads to significantly lower relative MILP gaps $\overline{\mathbf{\Delta}_s^{\text{lim}}}$ at the time limit and significantly more explored nodes $\overline{\mathbf{n}_s^{\text{nodes}}}$ and separated Benders cuts $\overline{\mathbf{n}_s^{\text{boc}}}$ for all newly defined strategies. Note that in $\overline{\mathbf{n}_s^{\text{boc}}}$, we only count the number of separated Benders cuts that comprise both Benders cuts added by

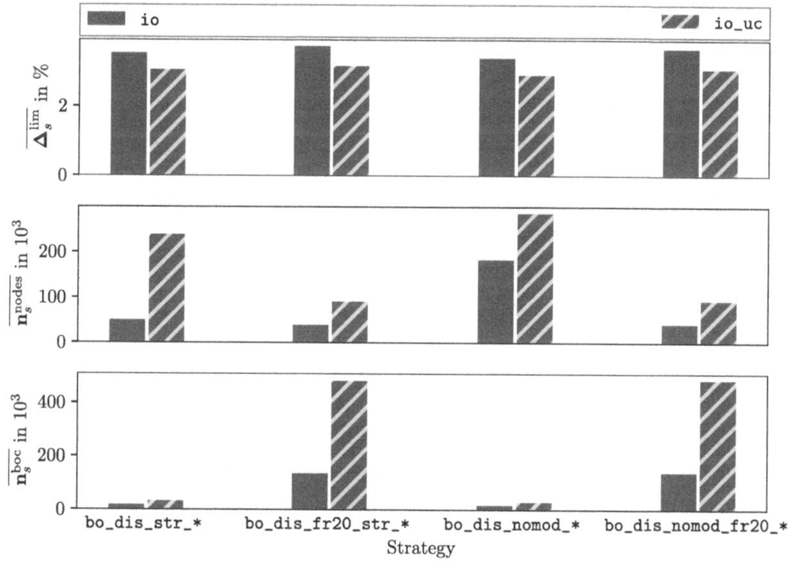

Fig. 6.7 Comparison of shifted geometric means of relative MILP gap at the time limit, number of explored nodes, and number of Benders optimality cuts in B&BC for strategies bo_dis_str_io_uc, bo_dis_fr20_str_io_uc, bo_dis_nomod_io_uc and bo_dis_nomod_fr20_io_uc as well as the respective strategies that do not add Benders cuts separated at fractional Benders RMP solutions as user cuts. The strategies optimize instances of 10b_20y. None of the instances is solved to proven optimality

means of lazy constraints and user cuts. The results indicate that adding Benders cuts separated at fractional Benders RMP solutions as user cuts help reduce the size of the Benders RMP and through the faster reoptimization facilitates the exploration of more nodes. For this reason, in the next computational study, we only consider strategies that utilize this technique.

Simplex Method for Optimization of Benders SPs

In this last computational study of this third stage, we change the simplex method with which we optimize our Benders SPs. As explained in Sect. 2.1.2, the dual simplex method can be started directly with a previously found optimal solution when fixing linking variables of the Benders SP to values of a new Benders RMP solution. When using the primal simplex method, the dual feasible basis needs to be converted into a primal feasible basis, producing computational overhead. However, since we do not incorporate a Benders cut selection method into our framework, the primal simplex method might find different optimal solutions and thus different Benders cuts.

To evaluate the effect of changing the simplex method, we define new strategies `bo_dis_str_io_uc_prim`, `bo_dis_fr20_str_io_uc_prim`, `bo_dis_nomod_io_uc_prim` and `bo_dis_nomod_fr20_io_uc_prim` that differ from `bo_dis_str_io_uc`, `bo_dis_fr20_str_io_uc`, `bo_dis_nomod_io_uc` and `bo_dis_nomod_fr20_io_uc` in the configuration change that they optimize Benders SPs by means of the primal simplex method. With newly defined strategies we optimize instances of `10b_20y`. None of the instances `10b_20y` is solved to proven optimality by any of the newly defined strategies.

Figure 6.8 compares the relative MILP gap $\overline{\mathbf{\Delta}_s^{\text{lim}}}$ at the time limit, the number of explored nodes $\overline{\mathbf{n}_s^{\text{nodes}}}$, the time spent in callbacks $\overline{\mathbf{t}_s^{\text{cb}}}$, and the number of separated Benders optimality cuts $\overline{\mathbf{n}_s^{\text{boc}}}$ of strategies `bo_dis_str_io_uc_prim`, `bo_dis_fr20_str_io_uc_prim`, `bo_dis_nomod_io_uc_prim` and `bo_dis_nomod_fr20_io_uc_prim` with the respective strategies that optimize Benders SPs using the dual simplex method. The relative MILP gap $\overline{\mathbf{\Delta}_s^{\text{lim}}}$ at the time limit only slightly decreases with strategies `bo_dis_str_io_uc_prim` and `bo_dis_fr20_str_io_uc_prim`, while it stagnates and slightly increases when optimizing with strategies `bo_dis_nomod_io_uc_prim` and `bo_dis_nomod_fr20_io_uc_prim`. Interestingly, optimizing Benders SPs using the primal simplex method leads to significantly more nodes explored with strategies `bo_dis_str_io_uc_prim`, `bo_dis_fr20_str_io_uc_prim`. The reason for this increase in $\overline{\mathbf{n}_s^{\text{nodes}}}$ could be Benders cuts that are on average stronger, due to the potential dual degeneracy of Benders SPs. Note that

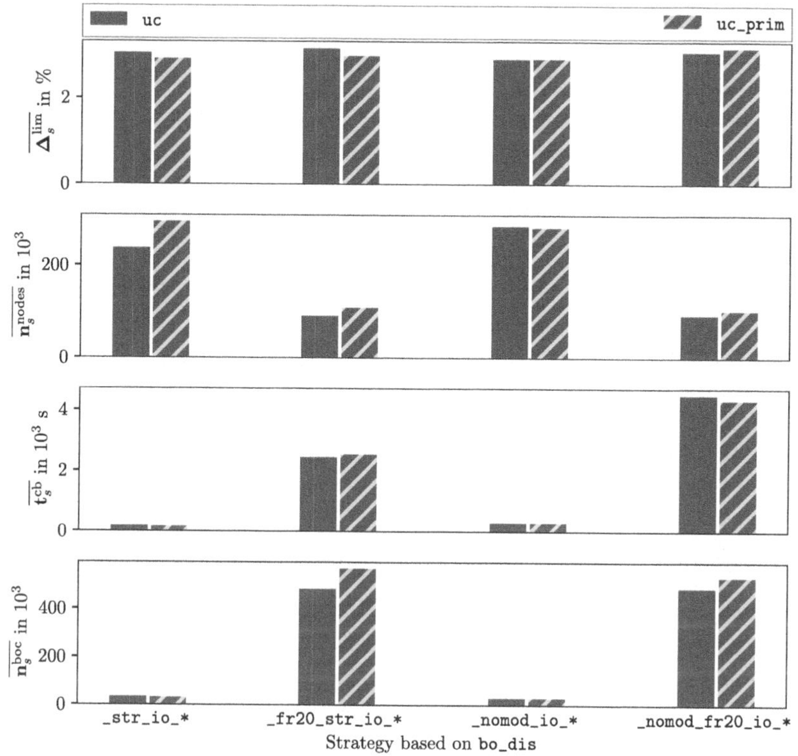

Fig. 6.8 Comparison of shifted geometric means of average MILP gap at termination, number of explored nodes, time in solver callbacks and number of Benders optimality cuts in B&BC for strategies `bo_dis_str_io_uc_prim`, `bo_dis_fr20_str_io_uc_prim`, `bo_dis_nomod_io_uc_prim` and `bo_dis_nomod_fr20_io_uc_prim` as well as the respective strategies that optimize Benders SPs by means of the dual simplex method. The strategies optimize instances of `10b_20y`. None of the instances is solved to proven optimality

using the primal simplex method does not significantly increase the time in solver callbacks for any of the considered strategies. However, we can observe a slight increase in the number of separated Benders optimality cuts $\overline{\mathbf{n}_s^{\text{boc}}}$ for strategies `bo_dis_fr20_str_io_uc_prim` and `bo_dis_nomod_fr20_io_uc_prim`. At this point, we note that the time spent in callbacks is low in comparison to the time spent solving the Benders RMP. Therefore, the impact of switching to the primal simplex method might be insubstantial since the optimization times of Benders SPs are relatively fast. In Benders decompositions with larger Benders SPs and thus longer reoptimization times, optimizing Benders SPs by the primal simplex method might significantly increase time spent in callbacks.

To evaluate the transferability of previous results and insights, we optimize instances of `20b_20y` with strategies `bo_dis_str_io_uc_prim`, `bo_dis_fr20_str_io_uc_prim`, `bo_dis_nomod_io_uc_prim` and `bo_dis_nomod_fr20_io_uc_prim`. None of the considered strategies solves any instance of `20b_20y` to proven optimality.

Table 6.11 shows $\overline{\mathbf{\Delta}_s^{\mathrm{lim}}}$, $\overline{\mathbf{\Delta}_s^{\mathrm{root}}}$, $\overline{\mathbf{t}_s^{\mathrm{root}}}$, $\overline{\mathbf{t}_s^{\mathrm{prim}}}$ and $\overline{\mathbf{n}_s^{\mathrm{nodes}}}$ of aforementioned strategies for the optimization of instances of `20b_20y`. Overall, relative MILP gaps $\overline{\mathbf{\Delta}_s^{\mathrm{lim}}}$ at the time limit are slightly better for strategies that only separate Benders cuts for fractional solutions at the root node, i.e., `bo_dis_nomod_io_uc_prim` and `bo_dis_str_io_uc_prim` compared to strategies that separate Benders cuts for fractional solutions at every 20th node of the tree. This is a consistent observation with the performance of the same strategies on instance set `10b_20y`. Here, strategy `bo_dis_nomod_io_uc_prim` reaches the lowest MILP gap at the time limit with 4 %, even though strategy `bo_dis_str_io_uc_prim` attains an only slightly larger relative MILP gap of 4.08 %. As expected, the relative MILP gaps for all strategies increased on average for the optimization of instances of `20b_20y`. However, the order between strategies for values of $\overline{\mathbf{\Delta}_s^{\mathrm{lim}}}$ is maintained. Therefore, in the next section, we choose strategy `bo_dis_nomod_io_uc_prim` to optimize instances in instance set `50b_20y`, assuming robustness of our computational results when optimizing larger instances than the ones in `20b_20y`.

Table 6.11 Relative MILP gap at the time limit, relative MILP gap when leaving the root node, time spent at the root node, time to determine the best primal bound and the number of explored nodes for strategies of the third stage for the optimization of instances in `20b_20y`. None of the instances is solved to proven optimality by any of the considered strategies

Strategy	$\overline{\mathbf{\Delta}_s^{\mathrm{lim}}}$ (%)	$\overline{\mathbf{\Delta}_s^{\mathrm{root}}}$ (%)	$\overline{\mathbf{t}_s^{\mathrm{root}}}$ (s)	$\overline{\mathbf{t}_s^{\mathrm{prim}}}$ (s)	$\overline{\mathbf{n}_s^{\mathrm{nodes}}}$
`bo_dis_str_io_uc_prim`	4.08	7.43	1190.79	25564.77	134637.66
`bo_dis_fr20_str_io_uc_prim`	4.30	7.39	1253.84	12522.76	54550.55
`bo_dis_nomod_io_uc_prim`	4.00	7.39	1237.11	26623.41	144098.74
`bo_dis_nomod_fr20_io_uc_prim`	4.41	7.38	1326.03	16020.31	61081.57

6.2.4 Optimization of Large Instances

Since it is of interest to optimize larger instance sizes to get a more accurate representation of the real building stock, we evaluate the possibility of optimizing instances of 50b_20y with $n^{\text{buil}} = 50$ buildings using the previously determined strategy bo_dis_nomod_io_uc_prim. Furthermore, we compare the performance of bo_dis_nomod_io_uc_prim against the performances of gurobi and gurobi_ws. For optimizations performed here, we use 16 CPUs and a time limit of 108 000 s, cf. Table 6.2, 6.3 and 6.4.

Table 6.12 shows $\overline{\mathbf{\Delta}_s^{\text{lim}}}$, $\overline{\mathbf{\Delta}_s^{\text{root}}}$, $\overline{\mathbf{t}_s^{\text{root}}}$, $\overline{\mathbf{t}_s^{\text{prim}}}$ and $\overline{\mathbf{n}_s^{\text{nodes}}}$ for strategies bo_dis_nomod_io_uc_prim, gurobi and gurobi_ws. None of the strategies solves any instance of instance set 50b_20y to optimality. On average, the relative MILP gap $\overline{\mathbf{\Delta}_s^{\text{lim}}}$ at the time limit is approximately cut in half between strategies gurobi and gurobi_ws as well as between gurobi_ws and bo_dis_nomod_io_uc_prim, reaching an average relative MILP gap of 5.15 % at the time limit. The use of the warm start solution significantly improves the relative MILP gap $\overline{\mathbf{\Delta}_s^{\text{lim}}}$ at the time limit for strategy gurobi_ws which is consistent with the observation in Sect. 6.2.1 for instances of 10b_20y and 20b_20y. Here, strategies gurobi and gurobi_ws spend the entire optimization time processing at the root node and thus do not start branching within the time limit. Conversely, strategy bo_dis_nomod_io_uc_prim spends on average 6227 s at the root node.

Table 6.12 Relative MILP gap at the time limit, relative MILP gap when leaving the root node, time spent at the root node, time to determine best primal bound and number of explored nodes for strategies optimizing instances of 50b_20y. None of the instances is solved to proven optimality by any of the considered strategies

Strategy	$\overline{\mathbf{\Delta}_s^{\text{lim}}}$ (%)	$\overline{\mathbf{\Delta}_s^{\text{root}}}$ (%)	$\overline{\mathbf{t}_s^{\text{root}}}$ (s)	$\overline{\mathbf{t}_s^{\text{prim}}}$ (s)	$\overline{\mathbf{n}_s^{\text{nodes}}}$
gurobi	20.07	7.86	104861.13	50460.99	0.00
gurobi_ws	10.01	7.84	104584.70	9989.00	0.00
bo_dis_nomod_io_uc_prim	5.15	8.75	6227.29	98210.86	138558.10

6.3 Discussion

In the following, we summarize the results of our computational studies and discuss the most relevant findings. Subsequently, we cover some of the limitations of our computational studies. Lastly, we deduce possible future research paths that build upon the solution approach developed in this work.

6.3.1 Summary

To conduct our computational studies, we first introduce a structure where we group changes to strategies based on model-based, algorithm-based, and solver-based configurations. Based on those three stages, we conduct different computational studies on instance sets `10b_20y` and `20b_20y`, while further refining strategies mainly based on relative MILP gap $\overline{\mathbf{\Delta}_s^{\mathrm{lim}}}$ at the time limit. After computational studies, we perform a last computational study optimizing instances of `50b_20y` with the best-performing strategy in terms of $\overline{\mathbf{\Delta}_s^{\mathrm{lim}}}$ on instances of `20b_20y`.

In computational studies at the first stage with instances of `10b_20y` and `20b_20y`, we find that base strategies `bo_base` and `bo_dis_base` based on decomposition BO reach significantly smaller relative MILP gaps $\overline{\mathbf{\Delta}_s^{\mathrm{lim}}}$ at the time limit than base strategies based on decompositions BOEP, BOESP, and BOESEP. Furthermore, results indicate that the further disaggregation of Benders SP and cuts by years $i \in I$ of decomposition BO reduces $\overline{\mathbf{\Delta}_s^{\mathrm{lim}}}$. The application of the warm start solution by our construction heuristic significantly improves $\overline{\mathbf{\Delta}_s^{\mathrm{lim}}}$ with strategy `gurobi_ws` in direct comparison with `gurobi`. Furthermore, we observe that the application of modified constraints reduces time spent in solver callbacks $\overline{\mathbf{t}_s^{\mathrm{cb}}}$. However, for decomposition BO, using modified constraints leads to slightly increased $\overline{\mathbf{\Delta}_s^{\mathrm{lim}}}$. The employment of valid inequalities for decompositions BOEP, BOESP, and BOESEP leads to the separation of Benders feasibility cuts and consistently increases the time spent in solver callbacks $\overline{\mathbf{t}_s^{\mathrm{cb}}}$. Lastly, we find that the best primal bounds in this stage are determined by strategy `bo_dis_base`, outperforming `gurobi_ws`.

At the second stage of our computational studies, we first separate Benders cuts for fractional solutions with $d^{\mathrm{nth}} = 20$ in strategies `bo_fr20`, `bo_dis_fr20` and `bo_dis_nomod_fr20`. The additionally separated Benders cuts only decrease $\overline{\mathbf{\Delta}_s^{\mathrm{lim}}}$ with `bo_fr20` that, however, still yields significantly larger values of $\overline{\mathbf{\Delta}_s^{\mathrm{lim}}}$ than strategies `bo_dis_fr20` and `bo_dis_nomod_fr20`. Therefore, for further computational studies, we decide to only further refine strategies `bo_dis_base`, `bo_dis_fr20`, `bo_dis_nomod` and `bo_dis_nomod_fr20`. Subsequently, we investigate the effect of strengthening Benders optimality cuts for strategies `bo_dis_base` and `bo_dis_fr20` with new strategies `bo_dis_str` and `bo_dis_fr20_str`. The strengthening of Benders optimality cuts significantly improves the best determined dual bounds $\underline{z}^{\text{lim}}_{p,s}$ and relative MILP gaps $\overline{\boldsymbol{\Delta}^{\text{lim}}_{s}}$ at the time limit. In a last computational study of the second stage, we investigate the effect of employing the in-out method with strategies `bo_dis_str_io`, `bo_dis_fr20_str_io`, `bo_dis_nomod_io` and `bo_dis_nomod_fr20_io`. The employment of the in-out method improves $\overline{\boldsymbol{\Delta}^{\text{lim}}_{s}}$ for all four strategies.

In computational studies at the third stage, we first evaluate the effect of adding separated Benders cuts for fractional Benders RMP solutions as user cuts instead of lazy cuts to the Benders RMP. For this, we define new strategies `bo_dis_str_io_uc`, `bo_dis_fr20_str_io_uc`, `bo_dis_nomod_io_uc` and `bo_dis_nomod_fr20_io_uc`. This procedure leads to significantly lower $\overline{\boldsymbol{\Delta}^{\text{lim}}_{s}}$, a larger number of explored nodes $\overline{\mathbf{n}^{\text{nodes}}_{s}}$ and separated Benders optimality cuts $\overline{\mathbf{n}^{\text{boc}}_{s}}$ for all four strategies. Furthermore, we observe that for each Benders SP included in the decomposition, exactly one Benders optimality cut is tight after the last iteration of the in-out method. In a last computational study, instead of optimizing Benders SPs with the dual simplex method, we optimize Benders SPs with the primal simplex method to get some further insights into the impact of potential dual degeneracy of Benders SPs. With newly defined strategies `bo_dis_str_io_uc_prim`, `bo_dis_fr20_str_io_uc_prim`, `bo_dis_nomod_io_uc_prim` and `bo_dis_nomod_fr20_io_uc_prim`, we observe that the optimization of Benders SPs with the primal simplex method slightly decreases $\overline{\boldsymbol{\Delta}^{\text{lim}}_{s}}$ in strategies `bo_dis_str_io_uc_prim` and `bo_dis_fr20_str_io_uc_prim` while $\overline{\boldsymbol{\Delta}^{\text{lim}}_{s}}$ stagnates and slightly increases for `bo_dis_nomod_fr20_io_uc_prim`, respectively. We then optimize instances in the instance set `20b_20y` with strategies s `bo_dis_str_io_uc_prim`, `bo_dis_fr20_str_io_uc_prim`, `bo_dis_nomod_io_uc_prim` and `bo_dis_nomod_fr20_io_uc_prim` to evaluate the transferability of gathered insights to larger instance sizes. While values of $\overline{\boldsymbol{\Delta}^{\text{lim}}_{s}}$ increase in comparison to the optimization of instances in `10b_20y` for the four strategies, the order between strategies for values of $\overline{\boldsymbol{\Delta}^{\text{lim}}_{s}}$ is maintained. The lowest value for $\overline{\boldsymbol{\Delta}^{\text{lim}}_{s}}$ for the optimization of instances in `20b_20y` is attained by strategy bo_dis_nomod_io_uc_prim that we therefore use for the following last computational study.

In a last computational study, we optimize instances of 50b_20y with strategies gurobi_ws, gurobi and bo_dis_nomod_io_uc_prim. The lowest relative MILP gap $\overline{\Delta_s^{\text{lim}}}$ at the time time limit is achieved by bo_dis_nomod_io_uc_prim with a value of 5.15 %. Strategies gurobi_ws and gurobi attain relative MILP gaps of 10.01. % and 20.07 %, respectively. Furthermore, in strategies gurobi_ws and gurobi, the solver does not leave the root node for any instance of 50b_20y until the time limit of 108 000 s.

6.3.2 Result Discussion

The computational results at the first stage indicate that the disaggregation of Benders SPs and Benders cuts in strategy bo_dis_base provides significantly stronger dual bounds than base strategies based on other Benders decompositions. While in Benders decompositions BOEP, BOESP, and BOESEP the corresponding Benders RMPs contain significantly fewer constraints than in decomposition BO, Benders cuts aggregated over years $i \in I$ are significantly weaker. Therefore, retaining the information that is projected out to Benders SPs by means of the model decomposition requires a large number of Benders cuts in decompositions BOEP, BOESP, and BOESEP. A possible explanation for this is that for a fixed building $b \in B$, there are possibly a very large number of feasible allocations of devices and efficiency measures over years $i \in I$. Therefore, if Benders cuts are aggregated by all years $i \in I$, small changes in the Benders RMP solution structure with respect to the building energy system device and building efficiency measure allocation over years $i \in I$ for that building could render previously separated Benders cuts useless.

We show that the modification of constraints reduces the time spent in solver callbacks. Depending on the specific Benders SPs, using fixing constraints and using continuous variables to replace linking variables or the RHS in constraints of the Benders SPs bears the potential to significantly reduce the time needed to generate Benders cuts. However, in our computational studies, the modification of constraints leads to different separated Benders cuts that perform worse than for strategies where we do not modify constraints in Benders SPs. A reason for this could be the dual degeneracy of Benders SP solutions. As explained in Sect. 2.3.4, the effect of dual degeneracy in solutions of Benders SPs is a common issue in Benders decomposition implementations.

The additional separation of Benders cuts for fractional Benders RMP solutions at node depths that are integer multiples of 20 does not significantly improve relative

MILP gaps at the time limit. Conversely, the additional separation of Benders cuts at those nodes leads to significantly more Benders cuts being added to the Benders RMP, slowing down the reoptimization time of the Benders RMP, and leading to fewer explored nodes. However, since the relative MILP gaps at the time limit do not significantly change, this could be an indication that the relative MILP gap only improves very little after some point during the B&BC process. This effect is commonly known as *tailing-off effect* (Rahmaniani et al., 2017). The application of the in-out method tightens the Benders RMP but does not add an excessive number of Benders cuts because of its cut removal mechanism. Strategies that employ the in-out method, therefore, have a faster reoptimization time of the Benders RMP and thus might explore more nodes, ending up with lower relative MILP gaps at the time limit.

Adding Benders cuts separated at fractional Benders RMP solutions as user cuts instead of lazy constraints improves the performance of B&BC. Without adding lazy constraints at fractional solutions, the Benders RMP does not grow excessively in size and thus, results show that more nodes can be explored, leading to lower relative MILP gaps. The observation that relative MILP gaps at the time limit are improved for some strategies when optimizing Benders SPs with the primal simplex method underscores the indication of potential dual degeneracy in dual solutions of Benders SPs.

The results of our computational studies indicate that the proposed solution approach is capable of effectively determining better dual and primal bounds than the commercial MILP solver Gurobi for instances of the introduced problem. Moreover, we expect this finding to hold for even larger instances than the ones in `50b_20y` as increasing the optimized instance size in our computational studies does not change this relation. Furthermore, we note that in most of the observed cases, the construction heuristic provides high-quality primal solutions. Therefore, it is conceivable that in a practical application of the proposed solution method to instances with large numbers of buildings, the construction heuristic is used to determine a primal solution and the following B&BC is conducted until a dual bound is attained for which the resulting relative MILP gap is satisfactory. Potentially, such dual bounds might be found relatively early during B&BC, due to the aforementioned tailing-off effect.

6.3.3 Limitations

We formulate the problem description and the corresponding MILP model in Chap. 3 based upon existing work in Richarz et al. (2022), Kotzur et al. (2020) and Schütz

(2018) (amongst others) and our expertise in energy engineering. In particular, we approximate the problem on the level of building operations by disregarding certain modeling characteristics such as ramps, minimum up- and down times of devices, part-load dependencies of device efficiency or minimum part loads (Kotzur et al., 2021). Those modeling characteristics impose further restrictions, potentially rendering allocations of devices and efficiency measures determined by the model in this work infeasible. Note that the framework based on Benders decomposition developed here could deal with cases of infeasibilities in Benders SPs by the separation of Benders feasibility cuts. Furthermore, note that the described problem does not allow the new connection of a building to district heating within the optimization time frame.

Our computational studies are subject to several limitations. We base our conclusions on the computational results of data sets for which we estimate several parameters. Especially the parametrization for the allocation of restricted resources, i.e., device and workforce supply and working hours needed for the installation of devices or realization of efficiency measures, are rough estimates that might profoundly influence the optimization performance of all strategies. Moreover, in all instances, we minimized only $-NPV$, i.e., we set $\phi = 1$ in objective function (3.55). The performance of our solution approach as well as the benchmark strategies `gurobi` and `gurobi_ws` might depend significantly on different objective functions. Furthermore, we did not tune parameters for benchmark strategies `gurobi` and `gurobi_ws`, potentially underestimating the performance of the solver with appropriately tuned parameters.

6.3.4 Future Research

The results of our computational studies indicate that the performance of our strategies significantly depends on the dual solutions for which Benders cuts are separated. In the existing implementation, the in-out method cannot be used to tighten Benders cuts during B&BC like in Maher (2021) because of the lack of necessary callback mechanisms in Gurobi 10.0. For this reason, the relative effect of additional Benders cut strengthening could be investigated by using solvers that have an in-out method implemented, e.g, SCIP. Alternatively, other Benders cut selection mechanisms like the one of Saharidis and Ierapetritou (2013) and Oliveira et al. (2014) that use a relative interior point could be investigated.

We note at this point that we, based on the work of Papadakos (2008), experimented with the separation of additional Benders cuts for approximated relative interior points determined during B&BC by computing convex combinations of subsequent integer solutions of Benders RMPs. Early results indicated, however, that this procedure diminishes the performance of the solution approach. A reason for this might be that Papadakos (2008) approximates relative interior points using the classical Benders decomposition algorithm introduced in Sect. 2.3.1. During B&BC, however, integer solutions might subsequently be found in different parts of the B&BC tree and thus be structurally very different, resulting in separation points that do not help improve the dual bound significantly with separated Benders cuts. This effect should therefore be considered when investigating other cut selection mechanisms that are based on approximated relative interior points or are so far only applied in the context of the classical Benders decomposition algorithm. Note that the times spent in callbacks are very short for strategies that are based on `bo_dis_base`. Therefore, spending more time in solver callbacks in exchange for potentially stronger Benders cuts bears the potential to efficiently further reduce relative MILP gaps at the time limit or even solve some of the introduced problem instances to proven optimality.

Lastly, future research could focus on determining accurate data for instances to evaluate the effect of estimated parameters on the performance of our solutions strategies and consider additional sensitivity analyses to assess the impact of parameter changes on performance, especially for parameters that describe the resource allocation at the building stock level. Furthermore, the convex combination of $-NPV$ and TGW in the objective function facilitates the computation of points on a Pareto curve by iteratively changing the parameter ϕ. In those single optimizations, previously determined optimal solutions could be used to warm start the B&BC process to optimize the instance with a now changed ϕ. Furthermore, the construction heuristic could be adapted to also consider TGW in the optimization process.

Conclusion 7

In this thesis, the problem of determining optimal transformation pathways for a residential building stock is defined together with a MILP model for the described problem. Furthermore, we present the four different Benders decompositions BO, BOEP, BOESP, and BOESEP of the proposed model. For the four decompositions, we develop constraint modifications that aim to reduce the time needed for Benders cut generation and introduce different valid inequalities for Benders RMPs. Furthermore, we develop a strengthening technique for Benders optimality cuts of to be applied in some of the Benders decompositions. We implement a solution approach that embeds the different Benders decompositions in a B&BC algorithm by separating Benders cuts in callbacks of the solver Gurobi. In our implementation, we also include an optional construction heuristic that determines a primal solution to the original model as well as an optional in-out method that we apply before starting B&BC. To test the solution approach, instances are constructed based on data for the German residential building sector and bundled in different instance sets. We then test the solution approach on the different instance sets by iteratively refining different configurations of the solution strategy. Lastly, we optimize larger instances to evaluate the transferability of previously gathered results and insights.

The problem of determining optimal transformation pathways for the building sector consists of three levels. On the first level, regionally restricted resources, e.g. building energy system devices and workforce for the installation of such devices or the realization of building efficiency measures, are allocated on the building stock level along spatial and temporal decision axes. The particular allocation of resources over years of the time horizon and buildings of the building stock then results in building efficiency and building energy system pathways on the second level which we collectively define as transformation pathways. Transformation pathways, therefore, describe the state of each building energy system and the building efficiency along the time axis for each building. On the third level, building operations are

R. Delorme, *Optimization of Transformation Pathways for the Residential Building Sector*, BestMasters, https://doi.org/10.1007/978-3-658-47466-9_7

optimized such that energy demands are met in every time step, where the specific energy demands depend on the yearly state of the building efficiency. Since transformation pathways couple consecutive years along the time axis, the problem needs to be holistically considered for the entire time horizon to determine exact solutions. The considered objective is to minimize a convex combination of the negative NPV and TGW. Based on the described problem, we formulate a MILP model that solves the described problem. At the building energy system level, the model considers different devices for the generation of thermal and electrical energy and allows for considering existing connections of individual buildings to district heating.

Based on the proposed MILP model, we apply Benders decomposition to decompose the original model into four different decompositions. The decompositions differ by which constraints are included in Benders SPs. While decomposition BO only includes constraints pertaining to building operations in Benders SPs, Benders SPs of decompositions BOEP and BOESP additionally contain constraints pertaining to the description of building efficiency and building energy system pathways, respectively. In decomposition BOESEP, Benders SPs contain all previously mentioned constraints. Decomposition BO differs from the other decompositions in the fact that its Benders SPs are further decomposable by years and Benders cuts can thus also be disaggregated by single years of optimization time horizon.

We introduce modified versions of all Benders SPs by modifying constraints such that linking variables in Benders SPs are replaced in constraints by a continuous copy and fixing this copy to the original value with additional constraints. Similarly, we replace RHSs unequal to zero with variables that are then fixed by additional constraints. This technique leads to larger Benders SPs but reduces the number of constraints to be checked for Benders cut generation. Furthermore, we introduce a technique to strengthen Benders optimality cuts for decompositions BO and BOESP. The strengthening technique is based on the property of the respective Benders SPs that for a given building efficiency state of a year, the optimal objective values of Benders SPs cannot improve through the choice of less efficient building efficiency states for any considered energy system design of that year. We informally name conditions for the parametrization of problem instances under which this strengthening technique can be applied. Lastly, for decompositions BOEP, BOESP, and BOESEP, we introduce valid inequalities that guarantee that only Benders optimality cuts may be separated during B&BC by retaining original restrictions on linking variables in the respective Benders RMP.

The decompositions are implemented in a B&BC framework using Python and callbacks of Gurobi 10.0. In the B&BC framework, Benders cuts are separated for integer solutions of the Benders RMP in the B&C process of Gurobi. Additionally, further Benders cuts can be separated at fractional Benders RMP solutions to strengthen the LP relaxation of the Benders RMP. For a given Benders RMP solution, Benders SPs are solved in parallel fashion to separate Benders cuts. We further implement an optional construction heuristic to compute an initial feasible solution to the original model with the aim to reduce the size of the B&BC tree and accelerate the optimization process. The construction heuristic determines an allocation of devices $u \in U_{\text{th}}^{\text{heat}}$ over years $i \in I$ while considering several constraints of the original model and valid inequalities developed for the Benders decompositions and is feasible if and only if the original model is also feasible. Moreover, we implement an optional in-out method that can be applied before B&BC to tighten the Benders RMP. In the in-out method, we iteratively add Benders cuts to an LP relaxation of the Benders RMP that we stabilize by a stabilization point that is updated at every iteration of the method. After a predefined number of iterations or when the method has converged, we remove Benders cuts from the LP relaxation that are not tight in the current solution.

For testing the implementation, we generate instances of a residential building stock based on data of the German residential building stock of the Census 2011 that statistically evaluates the residential building stock in terms of various properties, e.g., building geometry, location, or energy system. We categorize Germany into four regions and sample synthetic residential building stocks by assigning buildings to the different federal states that constitute the aforementioned regions. To explore different strategies by varying different configurations of our solution algorithm, we generate three instances sets `10b_20y`, `20b_20y` and `50b_20y` that respectively comprise 10, 20 and 50 buildings with an optimization time horizon of 20 years. To evaluate the performance of the different decompositions and the different implemented components, we choose to heuristically structure our computational studies in three stages pertaining to model-based, algorithm-based, and solver-based configurations to continuously refine our solution strategy. We conduct all computational studies of those three stages on instances of `10b_20y` or `20b_20y`. Eventually, we perform a computational study on instance set `50b_20y` with the solution strategy that shows the best performance in previous studies.

Computational studies on the first stage show that the best dual bounds are attained when using Benders decomposition BO with disaggregated Benders SPs and cuts. The relative MILP gaps at the time limit with this particular decomposition are consistently lower than when instances are solved directly with Gurobi, even if it is provided a warm start solution determined by our construction heuristic. Using the warm start solution significantly improves relative MILP gaps at the time limit when the original model is solved directly with Gurobi. While the employment of modified constraints lowers the time spent in solver callbacks, relative MILP gaps further decrease when not using modified constraints for decomposition BO with disaggregated Benders SPs and cuts. We identify the possible dual degeneracy of Benders SP solutions as a possible reason for this finding. The additional separation of Benders cuts for fractional Benders RMP solutions at node depths that are integer multiples of 20 leads to significantly fewer explored nodes but does not significantly change relative MILP gaps at the time limit. When subsequently strengthening Benders optimality cuts for strategies that employ decomposition BO with modified constraints, we find that dual bounds and relative MILP gaps at the time limit are significantly improved. The additional employment of the in-out method can further decrease relative MILP gaps at the time limit. In the third stage, we show that adding separated Benders cuts for fractional Benders RMP solutions as user cuts instead of lazy constraints considerably improves relative MILP gaps for all considered strategies. In a last computational study in the third stage, we optimize Benders SPs with the primal simplex method and show that for some strategies, the relative MILP gaps further decrease, supporting the hypothesis that dual degeneracy of Benders SP solutions leads to Benders cuts of different strength. Eventually, we select a best-performing strategy by comparing optimization results on instance set `20b_20y`. The best-performing strategy employs decomposition BO with disaggregated Benders SPs and cuts and unmodified constraints, uses a primal solution of our construction heuristic, only separates Benders cuts for fractional solutions at the root node and adds those as user cuts, employs the in-out method and optimizes Benders SPs with the primal simplex method. In our final computational study, we optimize instances of `50b_20y` with the aforementioned best-performing strategy and compare it against the performance of Gurobi with and without warm starting the optimization with a solution of the construction heuristic. In those optimizations, the strategy based on Benders decomposition attains relative MILP gaps of approximately 5 % at the time limit of the optimization. Conversely, the optimization with Gurobi yields relative MILP gaps of approximately 10 % and 20 % with and without performing a warm start, respectively.

The practical results of computational studies conducted in this thesis demonstrate the applicability of the solution approach to large-scale instances that are synthetically generated to test the implementation. Employing strategies based on Benders decomposition BO with disaggregated Benders SPs and cuts systematically provide better dual bounds than those to be found by Gurobi. Furthermore, we show that the construction heuristic is well suited to swiftly determine high-quality primal solutions if minimizing the negative NPV. Therefore, the developed solutions approach paves the way for the practical application for optimization of transformation pathways for residential building stock, given more accurate data is available.

Our computational studies are subject to several limitations. First, in the definition of the problem, we significantly simplify the characteristics of building energy system operations. The incorporation of more detailed characteristics of building energy system operations might render some solutions determined by our model infeasible. Moreover, we estimate several parameters for the generation of instances. Especially parameters that describe the allocation of restricted resources at the building stock level might significantly impact the performance of some strategies. Furthermore, we exploit certain parametrizations of problem instances for the application of our construction heuristic and the strengthening technique of Benders optimality cuts.

Future research could focus on building stock aggregation and the acquisition of more accurate and plausible data to deduct conclusions from optimization results beyond methodological considerations. Furthermore, the results of computational studies demonstrate that the performance of our solution approach could be further improved by employing an additional Benders cut selection scheme that deals with the potential degeneracy of dual solutions of Benders SPs. Moreover, the implementation itself could be further improved by adapting the construction heuristic to other values for ϕ than $\phi = 1$. Lastly, the computational results could be benchmarked against the performance of MILP solvers that employ general Benders decompositions frameworks, e.g., CPLEX or SCIP, to further evaluate the performance of the application-specific implementation.

Bibliography

Achterberg, Tobias (2008). *Constraint integer programming*. eng. 1. Aufl. Mathematik. Verl. Dr. Hut: München. ISBN: 978-3-89963-892-9.

Bazejczak, Jürgen and Dietmar Edler (2021). *Arbeitskräftebedarf nach Sektoren, Qualifikationen und Berufen zur Umsetzung der Investitionen für ein klimaneutrales Deutschland.* Kurzstudie im Auftrag der Bundestagsfraktion Bündnis 90/Die Grünen. Berlin. URL: https://www.gruene-bundestag.de/fileadmin/media/gruenebundestag_de/themen_az/klimaschutz/pdf/2105_Kurzstudie_Arbeitskra__ftebedarf_Klimaneutralitaet.pdf.

BDEW, Bundesverband der Energie- und Wasserwirtschaft e.V. (2019). *Studie: Wie heizt Deutschland?* URL: https://www.bdew.de/energie/studie-wie-heizt-deutschland/ (visited on 08/04/2023).

Ben-Ameur, Walid and José Neto (Jan. 2007). "Acceleration of cutting-plane and column generation algorithms: Applications to network design". en. In: *Networks* 49(1), pp. 3–17. ISSN: 00283045, 10970037. https://doi.org/10.1002/net.20137. URL: https://onlinelibrary.wiley.com/doi/10.1002/net.20137 (visited on 01/12/2023).

Benders, J. F. (Dec. 1962). "Partitioning procedures for solving mixed-variables programming problems". en. In: *Numerische Mathematik* 4(1), pp. 238–252. ISSN: 0029-599X, 0945–3245. https://doi.org/10.1007/BF01386316. URL: http://link.springer.com/10.1007/BF01386316 (visited on 01/19/2023).

Berthold, Timo and Gregor Hendel (May 2021). "Learning To Scale Mixed-Integer Programs". In: *Proceedings of the AAAI Conference on Artificial Intelligence* 35(5), pp. 3661–3668. ISSN: 2374-3468, 2159–5399. https://doi.org/10.1609/aaai.v35i5.16482. URL: https://ojs.aaai.org/index.php/AAAI/article/view/16482 (visited on 07/06/2023).

Birge, John R. and François V. Louveaux (Mar. 1988). "A multicut algorithm for two-stage stochastic linear programs". en. In: *European Journal of Operational Research* 34(3), pp. 384–392. ISSN: 03772217. https://doi.org/10.1016/0377-2217(88)90159-2. URL: https://linkinghub.elsevier.com/retrieve/pii/0377221788901592 (visited on 01/18/2023).

BMJ, Bundesministerium der Justiz (2019). *Federal Climate Change Act of 12 December 2019 (Federal Law Gazette I, p. 2513), as last amended by Article 1 of the Act of 18 August 2021 (Federal Law Gazette I, p. 3905).*

Bonami, Pierre, Domenico Salvagnin, and Andrea Tramontani (2020). "Implementing Automatic Benders Decomposition in a Modern MIP Solver". en. In: *Integer Programming and Combinatorial Optimization*. Ed. by Daniel Bienstock and Giacomo Zambelli. Vol. 12125.

R. Delorme, *Optimization of Transformation Pathways for the Residential Building Sector*, BestMasters, https://doi.org/10.1007/978-3-658-47466-9

Series Title: Lecture Notes in Computer Science. Springer International Publishing: Cham, pp. 78–90. ISBN: 978-3-030-45770-9 978-3-030-45771-6. https://doi.org/10.1007/978-3-030-45771-6_7. URL: http://link.springer.com/10.1007/978-3-030-45771-6_7 (visited on 01/12/2023).

Chermakani, Deepak Ponvel (2015). "Optimal Aggregation of Blocks into Subproblems in Linear-Programs with Block-Diagonal-Structure". In: Publisher: arXiv Version Number: 2. https://doi.org/10.48550/ARXIV.1507.05753. URL: https://arxiv.org/abs/1507.05753 (visited on 05/21/2023).

Chvátal, Vašek (1983). *Linear programming*. A Series of books in the mathematical sciences. W.H. Freeman: New York. ISBN: 978-0-7167-1195-7 978-0-7167-1587-0.

Codato, Gianni and Matteo Fischetti (Aug. 2006). "Combinatorial Benders' Cuts for Mixed-Integer Linear Programming". en. In: *Operations Research* 54(4), pp. 756–766. ISSN: 0030-364X, 1526–5463. https://doi.org/10.1287/opre.1060.0286. URL: http://pubsonline.informs.org/doi/10.1287/opre.1060.0286 (visited on 01/20/2023).

Contreras, Ivan, Jean-François Cordeau, and Gilbert Laporte (Dec. 2011). "Benders Decomposition for Large-Scale Uncapacitated Hub Location". en. In: *Operations Research* 59(6), pp. 1477–1490. ISSN: 0030-364X, 1526–5463. https://doi.org/10.1287/opre.1110.0965. URL: https://pubsonline.informs.org/doi/10.1287/opre.1110.0965 (visited on 08/09/2023).

dena, Deutsche Energie-Agentur GmbH (2021). *Aufbruch Klimaneutralität. Eine gesamtgesellschaftliche Aufgabe.* Abschlussbericht. Berlin.

Destatis, German Federal Statistical Office (2011). *Census 2011*. URL: https://ergebnisse2011.zensus2022.de/datenbank/online/ (visited on 08/04/2023).

Destatis, German Federal Statistical Office (Dec. 2022). *Umweltökonomische Gesamtrechnungen. Private Haushalt und Umwelt.* URL: https://www.destatis.de/DE/Themen/Gesellschaft-Umwelt/Umwelt/UGR/private-haushalte/Publikationen/Downloads/haushalte-umwelt-pdf-5851319.pdf?__blob=publicationFile (visited on 05/25/2023).

DWD, Deutscher Wetterdienst (2015). *Test Refence Year TRY 2015*. URL: https://www.dwd.de/DE/leistungen/testreferenzjahre/testreferenzjahre.html (visited on 10/05/2022).

Evins, Ralph (June 2013). "A review of computational optimisation methods applied to sustainable building design". en. In: *Renewable and Sustainable Energy Reviews* 22, pp. 230–245. ISSN: 13640321. https://doi.org/10.1016/j.rser.2013.02.004. URL: https://linkinghub.elsevier.com/retrieve/pii/S1364032113000920 (visited on 05/31/2023).

Fischetti, Matteo, Ivana Ljubić, and Markus Sinnl (July 2017). "Redesigning Benders Decomposition for Large-Scale Facility Location". en. In: *Management Science* 63(7), pp. 2146–2162. ISSN: 0025-1909, 1526-5501. https://doi.org/10.1287/mnsc.2016.2461. URL: http://pubsonline.informs.org/doi/10.1287/mnsc.2016.2461 (visited on 01/12/2023).

Gabrielli, Laura and Aurora Greta Ruggeri (Nov. 2019). "Developing a model for energy retrofit in large building portfolios: Energy assessment, optimization and uncertainty". en. In: *Energy and Buildings* 202, p. 109356. ISSN: 03787788. https://doi.org/10.1016/j.enbuild.2019.109356. URL: https://linkinghub.elsevier.com/retrieve/pii/S0378778819301859 (visited on 06/02/2023).

Geoffrion, Arthur M. (July 1970). "Elements of Large Scale Mathematical Programming Part II: Synthesis of Algorithms and Bibliography". en. In: *Management Science* 16(11), pp. 676–691. ISSN: 0025-1909, 1526–5501. https://doi.org/10.1287/mnsc.16.11.676. URL: https://pubsonline.informs.org/doi/10.1287/mnsc.16.11.676 (visited on 05/30/2023).

Gurobi Optimization, LLC (2020). *User MIP start did not produce a new incumbent solution/ Another try with MIP start*. URL: https://support.gurobi.com/hc/en-us/community/posts/360073235912-User-MIP-start-did-not-producc-a-new-incumbent-solution-Another-try-with-MIP-start (visited on 08/03/2023).

Gurobi Optimization, LLC (2023). *Gurobi Optimizer Reference Manual*. URL: https://www.gurobi.com.

Heijungs, Reinout (Mar. 2014). "Ten easy lessons for good communication of LCA". en. In: *The International Journal of Life Cycle Assessment* 19(3), pp. 473–476. ISSN: 0948-3349, 1614–7502. https://doi.org/10.1007/s11367-013-0662-5. URL: http://link.springer.com/10.1007/s11367-013-0662-5 (visited on 04/13/2023).

ISE, Fraunhofer-Institut für Solare Energiesysteme (2022). *synPRO Load Profiles*. URL: https://synpro-lastprofile.de/gratis-profile/ (visited on 10/05/2022).

Jenabi, M., S. M. T. Fatemi Ghomi, S. A. Torabi, and S. H. Hosseinian (Dec. 2015). "Acceleration strategies of Benders decomposition for the security constraints power system expansion planning". en. In: *Annals of Operations Research* 235(1), pp. 337–369. ISSN: 0254-5330, 1572–9338. https://doi.org/10.1007/s10479-015-1983-5. URL: http://link.springer.com/10.1007/s10479-015-1983-5 (visited on 08/09/2023).

Kelley Jr., J. E. (Dec. 1960). "The Cutting-Plane Method for Solving Convex Programs". en. In: *Journal of the Society for Industrial and Applied Mathematics* 8(4), pp. 703–712. ISSN: 0368-4245, 2168–3484. https://doi.org/10.1137/0108053. URL: http://epubs.siam.org/doi/10.1137/0108053 (visited on 01/30/2023).

Kotzur, Leander (2018). *Future grid load of the residential building sector*. en. Vol. RWTH Aachen University. Artwork Size: pages 1 Online-Ressource (xxi, 213 Seiten) : Illustrationen, Diagramme, Karten Pages: pages 1 Online-Ressource (xxi, 213 Seiten) : Illustrationen, Diagramme, Karten Publication Title: Dissertation. RWTH Aachen University. https://doi.org/10.18154/RWTH-2018-231872. URL: http://publications.rwth-aachen.de/record/752116 (visited on 11/17/2022).

Kotzur, Leander, Peter Markewitz, Martin Robinius, Gonçalo Cardoso, Peter Stenzel, Miguel Heleno, and Detlef Stolten (Feb. 2020). "Bottom-up energy supply optimization of a national building stock". en. In: *Energy and Buildings* 209, p. 109667. ISSN: 03787788. https://doi.org/10.1016/j.enbuild.2019.109667. URL: https://linkinghub.elsevier.com/retrieve/pii/S0378778819319863 (visited on 11/17/2022).

Kotzur, Leander et al. (Nov. 2021). "A modeler's guide to handle complexity in energy systems optimization". en. In: *Advances in Applied Energy* 4, p. 100063. ISSN: 26667924. https://doi.org/10.1016/j.adapen.2021.100063. URL: https://linkinghub.elsevier.com/retrieve/pii/S266679242100055X (visited on 11/17/2022).

Lueddeckens, S., P. Saling, and E. Guenther (Apr. 2022). "Discounting and life cycle assessment: a distorting measure in assessments, a reasonable instrument for decisions". en. In: *International Journal of Environmental Science and Technology* 19(4), pp. 2961–2972. ISSN: 1735-1472, 1735–2630. https://doi.org/10.1007/s13762-021-03426-8. URL: https://link.springer.com/10.1007/s13762-021-03426-8 (visited on 04/27/2023).

Lutz, Christian, Lisa Becker, and Ulrike Lehr (2018). *Mögliche Engpässe für die Energiewende*. Tech. rep. Osnabrück. URL: https://www.bmwk.de/Redaktion/DE/Publikationen/Studien/moegliche-engpaesse-fuer-die-energiewende.pdf?__blob=publicationFile&v=8.

Magnanti, T. L. and R. T. Wong (June 1981). "Accelerating Benders Decomposition: Algorithmic Enhancement and Model Selection Criteria". en. In: *Operations Research* 29(3), pp. 464–484. ISSN: 0030-364X, 1526–5463. https://doi.org/10.1287/opre.29.3.464. URL: http://pubsonline.informs.org/doi/10.1287/opre.29.3.464 (visited on 01/18/2023).

Maher, Stephen J. (Apr. 2021). "Implementing the branch-and-cut approach for a general purpose Benders' decomposition framework". en. In: *European Journal of Operational Research* 290(2), pp. 479–498. ISSN: 03772217. https://doi.org/10.1016/j.ejor.2020.08.037. URL: https://linkinghub.elsevier.com/retrieve/pii/S0377221720307505 (visited on 01/12/2023).

McDaniel, Dale and Mike Devine (Nov. 1977). "A Modified Benders' Partitioning Algorithm for Mixed Integer Programming". en. In: *Management Science* 24(3), pp. 312–319. ISSN: 0025-1909, 1526–5501. https://doi.org/10.1287/mnsc.24.3.312. URL: http://pubsonline.informs.org/doi/10.1287/mnsc.24.3.312 (visited on 02/08/2023).

Nemhauser, George L. and Laurence A. Wolsey (1988). *Integer and combinatorial optimization*. Wiley-Interscience series in discrete mathematics and optimization. Wiley: New York. ISBN: 978-0-471-82819-8.

Oliveira, F., I.E. Grossmann, and S. Hamacher (Sept. 2014). "Accelerating Benders stochastic decomposition for the optimization under uncertainty of the petroleum product supply chain". en. In: *Computers & Operations Research* 49, pp. 47–58. ISSN: 03050548. https://doi.org/10.1016/j.cor.2014.03.021. URL: https://linkinghub.elsevier.com/retrieve/pii/S0305054814000768 (visited on 01/18/2023).

Pannier, Marie-Lise, Thomas Recht, Maxime Robillart, Patrick Schalbart, Bruno Peuportier, and Laurent Mora (Nov. 2021). "Identifying optimal renovation schedules for building portfolios: Application in a social housing context under multi-year funding constraints". en. In: *Energy and Buildings* 250, p. 111290. ISSN: 03787788. https://doi.org/10.1016/j.enbuild.2021.111290. URL: https://linkinghub.elsevier.com/retrieve/pii/S0378778821005740 (visited on 03/28/2023).

Papadakos, Nikolaos (July 2008). "Practical enhancements to the Magnanti–Wong method". en. In: *Operations Research Letters* 36(4), pp. 444–449. ISSN: 01676377. https://doi.org/10.1016/j.orl.2008.01.005. URL: https://linkinghub.elsevier.com/retrieve/pii/S0167637708000102 (visited on 01/12/2023).

Petkov, Ivalin, Alicia Lerbinger, Georgios Mavromatidis, Christof Knoeri, and Volker H. Hoffmann (May 2023). "Decarbonizing real estate portfolios considering optimal retrofit investment and policy conditions to 2050". en. In: *iScience* 26(5), p. 106619. ISSN: 25890042. https://doi.org/10.1016/j.isci.2023.106619. URL: https://linkinghub.elsevier.com/retrieve/pii/S258900422300696X (visited on 05/30/2023).

Petkov, Ivalin, Georgios Mavromatidis, Christof Knoeri, James Allan, and Volker H. Hoffmann (May 2022). "MANGOret: An optimization framework for the long-term investment planning of building multi-energy system and envelope retrofits". en. In: *Applied Energy* 314, p. 118901. ISSN: 03062619. https://doi.org/10.1016/j.apenergy.2022.118901. URL: https://linkinghub.elsevier.com/retrieve/pii/S0306261922003269 (visited on 05/31/2023).

Python Software Foundation (2023). *Python 3.10.11 Documentation—The Python Standard Library -multiprocessing—Process-based parallelism*. URL: https://docs.python.org/3.10/library/multiprocessing.html (visited on 05/10/2023).

Rahmaniani, Ragheb, Teodor Gabriel Crainic, Michel Gendreau, and Walter Rei (June 2017). "The Benders decomposition algorithm: A literature review". en. In: *European Journal of Operational Research* 259(3), pp. 801–817. ISSN: 03772217. https://doi.org/10.1016/j.

ejor.2016.12.005. URL: https://linkinghub.elsevier.com/retrieve/pii/S0377221716310244 (visited on 12/01/2022).

Richarz, Jan, Sarah Henn, Tanja Osterhage, and Dirk Müller (Jan. 2022). "Optimal scheduling of modernization measures for typical non-residential buildings". en. In: *Energy* 238, p. 121871. ISSN: 03605442. https://doi.org/10.1016/j.energy.2021.121871. URL: https://linkinghub.elsevier.com/retrieve/pii/S0360544221021198 (visited on 09/01/2022).

Ruggeri, Aurora Greta, Laura Gabrielli, and Massimiliano Scarpa (Sept. 2020). "Energy Retrofit in European Building Portfolios: A Review of Five Key Aspects". en. In: *Sustainability* 12(18), p. 7465. ISSN: 2071–1050. https://doi.org/10.3390/su12187465. URL: https://www.mdpi.com/2071-1050/12/18/7465 (visited on 08/08/2023).

Sá, Elisangela Martins de, Ricardo Saraiva de Camargo, and Gilberto de Miranda (Apr. 2013). "An improved Benders decomposition algorithm for the tree of hubs location problem". en. In: *European Journal of Operational Research* 226(2), pp. 185–202. ISSN: 03772217. https://doi.org/10.1016/j.ejor.2012.10.051. URL: https://linkinghub.elsevier.com/retrieve/pii/S0377221712008211 (visited on 01/25/2023).

Saharidis, Georgios K. D. and Marianthi G. Ierapetritou (Nov. 2013). "Speed-up Benders decomposition using maximum density cut (MDC) generation". en. In: *Annals of Operations Research* 210(1), pp. 101–123. ISSN: 0254-5330, 1572–9338. https://doi.org/10.1007/s10479-012-1237-8. URL: http://link.springer.com/10.1007/s10479-012-1237-8 (visited on 01/24/2023).

Santoso, Tjendera, Shabbir Ahmed, Marc Goetschalckx, and Alexander Shapiro (Nov. 2005). "A stochastic programming approach for supply chain network design under uncertainty". en. In: *European Journal of Operational Research* 167(1), pp. 96–115. ISSN: 03772217. https://doi.org/10.1016/j.ejor.2004.01.046. URL: https://linkinghub.elsevier.com/retrieve/pii/S0377221704002292 (visited on 01/12/2023).

Schrijver, Alexander (2011). *Theory of linear and integer programming*. eng. Nachdr. Wiley-Interscience series in discrete mathematics and optimization. Wiley: Chichester Weinheim. ISBN: 978-0-471-98232-6.

Schütz, Thomas (2018). "Optimal design of energy conversion units and building envelopes for residential neighborhoods". en. Artwork Size: pages 1 Online-Ressource (XXII, 140 Seiten) : Illustrationen, Diagramme Pages: pages 1 Online-Ressource (XXII, 140 Seiten) : Illustrationen, Diagramme Publication Title: Dissertation Volume: Rheinisch-Westfälische Technische Hochschule Aachen. PhD thesis. RWTH Aachen University. https://doi.org/10.18154/RWTH-2019-02124. URL: http://publications.rwth-aachen.de/record/755957 (visited on 04/09/2023).

Schütz, Thomas, Markus Hans Schraven, Marcus Fuchs, Peter Remmen, and Dirk Müller (Dec. 2018). "Comparison of clustering algorithms for the selection of typical demand days for energy system synthesis". en. In: *Renewable Energy* 129, pp. 570–582. ISSN: 09601481. https://doi.org/10.1016/j.renene.2018.06.028. URL: https://linkinghub.elsevier.com/retrieve/pii/S0960148118306591 (visited on 03/29/2023).

Sherali, Hanif D. and Brian J. Lunday (Nov. 2013). "On generating maximal nondominated Benders cuts". en. In: *Annals of Operations Research* 210(1), pp. 57–72. ISSN: 0254-5330, 1572–9338. https://doi.org/10.1007/s10479-011-0883-6. URL: http://link.springer.com/10.1007/s10479-011-0883-6 (visited on 01/18/2023).

Stursberg, Paul Melvin (2019). "On the Mathematics of Energy System Optimization". en. PhD Thesis. Technische Universität München.

Taillandier, Franck, Christophe Fernandez, and Amadou Ndiaye (Mar. 2017). "Real Estate Property Maintenance Optimization Based on Multiobjective Multidimensional Knapsack Problem: Real estate property maintenance optimization". en. In: *Computer-Aided Civil and Infrastructure Engineering* 32(3), pp. 227–251. ISSN: 10939687. https://doi.org/10.1111/mice.12246. URL: https://onlinelibrary.wiley.com/doi/10.1111/mice.12246 (visited on 06/02/2023).

Teichgraeber, Holger and Adam R. Brandt (Apr. 2022). "Time-series aggregation for the optimization of energy systems: Goals, challenges, approaches, and opportunities". en. In: *Renewable and Sustainable Energy Reviews* 157, p. 111984. ISSN: 13640321. https://doi.org/10.1016/j.rser.2021.111984. URL: https://linkinghub.elsevier.com/retrieve/pii/S1364032121012478 (visited on 03/29/2023).

UBA, Umweltbundesamt (2019). *Wohnen und Sanieren. Empirische Wohngebäudedaten seit 2002.* Hintergrundbericht. Dessau-Roßlau. URL: https://www.umweltbundesamt.de/sites/default/files/medien/1410/publikationen/2019-05-23_cc_22-2019_wohnenundsanieren_hintergrundbericht.pdf.

UBA, Umweltbundesamt, ed. (2022). *Berichterstattung unter der Klimarahmenkonvention der Vereinten Nationen und dem Kyoto-Protokoll 2022. Nationaler Inventarbericht zum Deutschen Treibhausgasinventar 1990–2020.* URL: https://www.umweltbundesamt.de/sites/default/files/medien/1410/publikationen/2022-05-31_climate-change_24-2022_nir-2022_de.pdf (visited on 05/25/2023).

Wolsey, Laurence A. (Sept. 2008). "Mixed Integer Programming". en. In: *Wiley Encyclopedia of Computer Science and Engineering*. Ed. by Benjamin W. Wah. John Wiley & Sons, Inc.: Hoboken, NJ, USA, ecse244. ISBN: 978-0-470-05011-8. https://doi.org/10.1002/9780470050118.ecse244. URL: https://onlinelibrary.wiley.com/doi/10.1002/9780470050118.ecse244 (visited on 05/22/2023).

Wolsey, Laurence A. (2021). *Integer programming*. Second edition. Wiley: Hoboken, NJ. ISBNISBN: 978-1-119-60653-6.